MINDING THE CLIMATE

Minding the Climate

How Neuroscience Can Help Solve Our Environmental Crisis

ANN-CHRISTINE DUHAIME, MD

Harvard University Press
Cambridge, Massachusetts | London, England | 2022

Printed in the United States of America

First printing

Library of Congress Cataloging-in-Publication Data

Names: Duhaime, Ann-Christine, author.
Title: Minding the climate : how neuroscience can help solve our environmental crisis / Ann-Christine Duhaime, MD.
Description: Cambridge, Massachusetts : Harvard University Press, 2022. | Includes bibliographical references and index.
Identifiers: LCCN 2021062984 | ISBN 9780674247727 (cloth)
Subjects: LCSH: Neuropsychology. | Brain—Psychological aspects. | Climate change mitigation. | Brain—Evolution. | Neural circuitry—Adaptation.
Classification: LCC QP360 .D685 2022 | DDC 612.8—dc23/eng/20220316
LC record available at https://lccn.loc.gov/2021062984

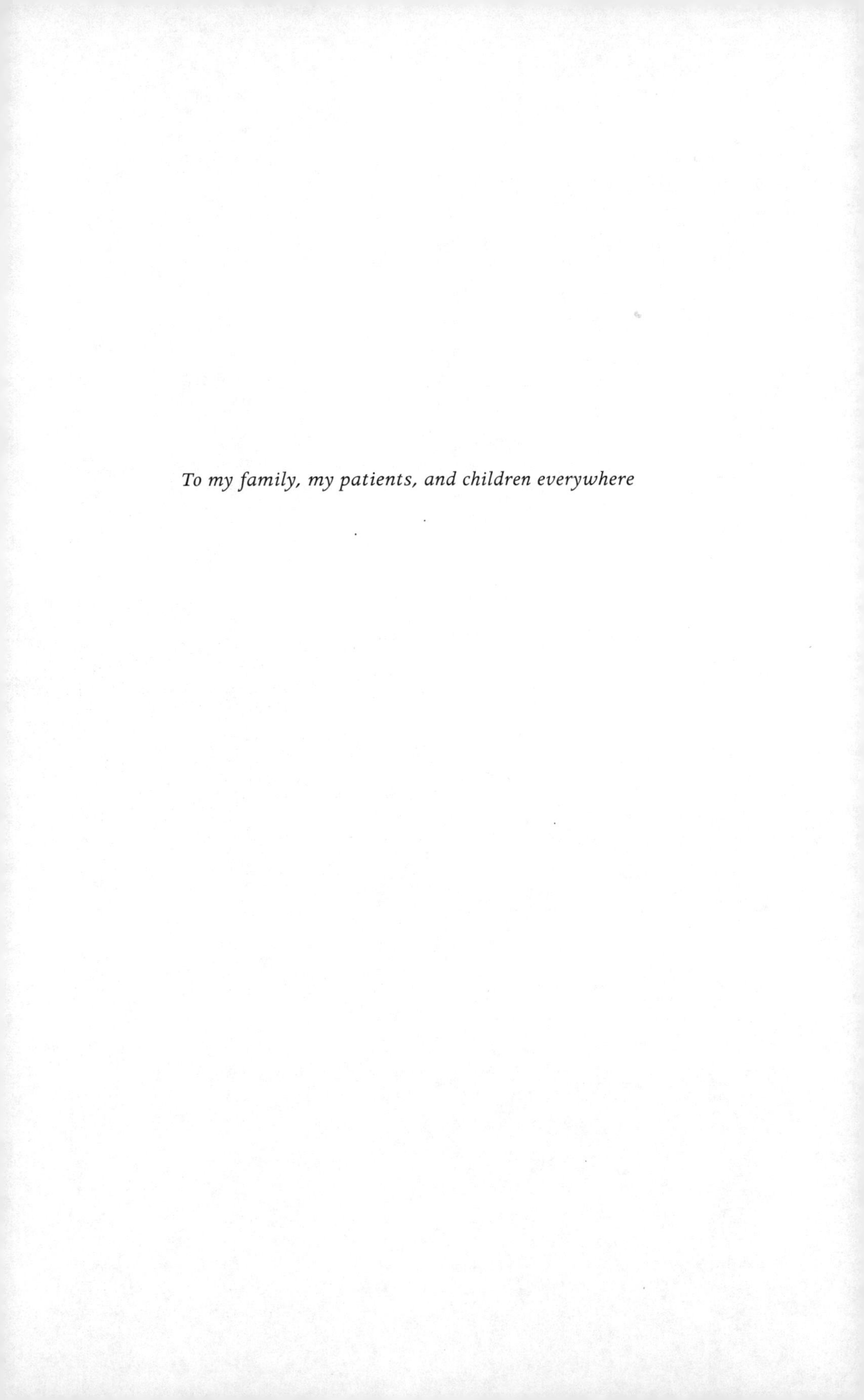

To my family, my patients, and children everywhere

Contents

Preface

It's 8 P.M. and I'm just about to pack up and leave the hospital for the night when the pager bleats. It's my resident, the junior neurosurgeon-in-training on for emergency calls, who quickly summarizes an urgent consult from the pediatric intensive care unit (PICU). "It's a six-year-old. He's on his third liver transplant for a congenital disease, and his blood clotting is way off. His scan shows a 5-centimeter left brain hemorrhage with shift, and he's comatose, with a dilated left pupil and a dropping heart rate. They want us to operate." I ask if the PICU team has corrected his blood clotting and am told they've given him factor, and he should be good to go to surgery. I join the resident in the operating room after stopping briefly to meet the family and explain as gently as possible that while we'll do our best, this is a grave situation. We may not be able to control the hemorrhage, and their child may not survive the surgery. They nod numbly and cry quietly. After months in the hospital, they can't do much more.

In surgery, after we suction out the bulk of the clot, the brain continues to ooze from every surface we've touched. I ask the circulating nurse to call the transplant attending and ask him if there's anything else they can do medically to help the blood to clot. While she holds the phone to my ear, I am told they've given appropriate correctives, and the bleeding should be easy to control. I invite them to come look, because it isn't easy to control—in fact, it's impossible to control. After using every hemostatic agent we have in the hospital and every kind of transfusion that might be helpful and every way to apply gentle pressure to the welling surfaces, we finally recognize that we can only slow, not stop, this hemorrhage. We will have to close up enough to get him back to the PICU and wait. He probably will not survive the night.

Other children fare better. Their seizures are localized with magnetoencephalography and suppressed with minimally invasive techniques, their tumors are resected with help from computerized stereotactic image guidance or laser ablation, their infections successfully overcome with fifth- and sixth-generation designer antibiotics. They pass in and out of our medical lives, but the traces they leave pale in comparison to those of the children we can't help. Both the successful and the futile treatments require enormous resources in our advanced medical system. New machines, tests, and technologies are added every day, requiring specialized spaces, protocols, and people. We are trained to treat the child before us and not to think of the resources or the energy it takes; if something exists that might help, our job is to think of it and to use it. We must not be distracted by larger issues of who gets what.

Meanwhile, on other continents, desperate families risk everything to flee civil wars made worse by a decade of drought. In places where population growth is exploding, the average-age citizen is too young to vote but not too young to suffer when flooding occurs and disease follows, and entire extended families are wiped out. Others live in economic and spiritual poverty, because traditional ways of life and the rituals and society that sustained them are being swept away by physical alteration of a human-changed planet. Closer to home, fires and floods and droughts wreak havoc; plastic litters everything. Those who already are the most disadvantaged bear the biggest burden, as power plants and polluting industries disproportionately sully their neighborhoods. Parents everywhere worry about pesticides and toxins and rising temperatures. But we are the good guys. Our hospital produces cutting-edge therapies and daily small miracles along with enormous quantities of carbon dioxide, trash, and poisonous waste that threaten the very futures we've worked to make possible.

Mostly, we compartmentalize these contradictions. Our brains allow for this.

The American Academy of Pediatrics has stated that climate change is the biggest public health problem we face as a species. This affects all of us and will affect people all around the world—first, the most vulnerable, and children in particular. But while we know this intellectually, and it makes us feel bad, and often helpless, it doesn't typically influence our day-to-day behavior. The conflict between this truly "wicked problem"—one that is complicated, involving many intersecting and contradictory features and without a simple solution—and our values and actions in our daily tasks was the impetus for this book.

As most health-care providers who care for children can attest, it is an extraordinary privilege and is tremendously rewarding to have the opportunity to help patients and their families through medical crises. I chose my career at a time before climate change was clearly recognized, but even then, the problems of exponential population growth and environmental threats seemed to me to be the most consequential ones to tackle. The interdependence of all life forms seems an obvious fact of biology, and we ignore it at our peril. Yet I couldn't escape the hypnotic pull of fascination with the brain—the means and mechanisms of why we are who we are. Like most neurosurgeons, I can remember vividly the first patient, the first case, the moment I was "hooked." No other career could allow me to experience the utter amazement of seeing the human brain in action, up close and tangible, with the chance to influence someone's life for the better by quelling a threat to this most awesome invention of nature—the very essence of a person's existence.

Just as engineers and artists and cooks perceive the world filtered through what is on their minds most of the time, neurosurgeons are immersed in the brain—how it's designed, how it develops, how it is injured and repaired, how it interacts with the world to create and modify every unique person through all the phases and changes of life. We observe these things in our patients recovering from injuries or surgery, from watching them grow up as we follow them for chronic conditions, and I saw them in my laboratory research on brain recovery and plasticity. From the perspective of many of us in the neuroscience fields, *everything* ultimately is about the brain. How we act, what we choose, what we value, the very history of our species with all its accelerating change reflecting advances in science and technology—it all can be explained through the lens of how our brains work.

The brain, like everything else in biology, was designed under the tutelage of evolution. It follows certain principles and has strong tendencies but is also exquisitely responsive to the culture and circumstances in which it finds itself. We find it rewarding to care for one child at a time, and to bring to bear all we are able toward our success, because we are designed that way. Caring for another human being who is right in front of us, expending effort and skill to make that person better, reflects some of the ways we evolved through millennia to function. But when considering larger social issues, it didn't seem surprising to me that climate change—arguably the most important problem for us as a species—was a particularly difficult one for us to address. It's easy to understand its causes through a neural lens—our extraordinary brains equipped us to adapt, expand, and change the world to our immediate benefit, spurred by a design favoring decision making based on short-term consequences. But fixing climate change is harder; it's just not the kind of problem our brains evolved to perceive or solve easily. In fact, the choices most needed right now to curb the worst trajectories of climate change often run directly contrary to the way we "work" brain-wise.

In 1998, environmental writer and advocate Bill McKibben wrote a three-part series for *Atlantic* magazine, "A Special Moment in History," that was part of a look ahead at the biggest questions of the new millennium.[1] McKibben noted, with some surprise, that changing behavior relative to population seems easier than changing behavior relative to our incessant consumption, which is the root cause of the skyrocketing levels of atmospheric carbon dioxide that invisibly heats our planet. He posited that we just wouldn't be able to live simply enough, soon enough, to reverse this trend. Of course, I thought. Simplification is difficult for humans. It's not the way our brains work. If we don't understand that root cause, it will be difficult to change in the right direction. I mulled over these thoughts, uncertain about how they might translate into any kind of useful action.

The weight of the contradiction between the looming crisis of climate change and the intensity demanded by my "one child at a time" career and high-resource medical culture became sufficiently uncomfortable that by the middle of my career, doing *something* became imperative. As people concerned about health, how could we ignore this, the problem that threatened everything, and begged for action? The words of my wise mentor in pediatric neurosurgery, Dr. Luis Schut, who had traveled on humanitarian missions to all corners of the world, bored into my consciousness throughout my second decade in the field: "If all the neurosurgeons in the world went up in a puff of smoke tomorrow, the change would amount to but one drop in the vast ocean of human suffering." He told us this when we veered toward getting too big for our britches, to remind us that this rewarding career, while challenging and of utmost importance to the specific children and families we served, was not a big-picture enterprise.

Of course, no individual alone can solve these complex, global problems, including the resource inequities that underlie so much of human misery. When I sought the advice of the chair of the environmental science department at my alma mater, Brown University, as to what I might do to help in this urgent sphere of climate change, his response was simple: Your best chance of having an impact is to work within your field. Would exploring the interrelations between brain reward and our stubborn reluctance to engage in pro-environmental behavior offer any practical insights that might help us change more easily? The opportunity came by sheer good fortune when I was awarded a one-year fellowship at the Radcliffe Institute for Advanced Study at Harvard University, where I could pursue these ideas while relieved temporarily of day-to-day responsibilities for patient care. Radcliffe offered a cauldron of incredible people from a wide variety of fields and perspectives, and this interdisciplinary free-for-all was an invaluable part of the process of exploring and refining these ideas. It amazed me that even some exceptionally intelligent and knowledgeable people, experts in areas like advanced computational science, didn't accept or fully understand that a two-degree change in average planetary temperature could be meaningful or important. This was eye-opening, and it led to spirited discussions between those who studied environment-related issues and those who knew less about them and didn't fully grasp the implications. We all had a lot to learn.

Minding the Climate, the result of this exploration, addresses a series of questions about what evidence exists for the brain / environment connection, and what implications this connection has for pro-environmental action. The starting point for this journey arose from my own early research using deep brain stimulation to tease apart the workings of the brain reward system. My later research, focused on brain development, plasticity, and repair, contributed additional dimensions. More recent immersion in creative scientific work by my colleagues on the detailed neural circuitry of the reward system and its role in decision making, as well as psychiatric and behavioral disorders relevant to consumption, offered new perspectives on how the brain's evolutionarily guided design might indeed be relevant to our environmental crisis.

From its starting point in neuroscience, the path of this exploration extended to overlapping areas in evolutionary biology, psychology, economics, consumer studies, marketing, sociology, public health, child development, education, environmental science, and policy. In exploring connections to other disciplines,

I drew primarily on research published in peer-reviewed sources, as these studies would provide the most objectively verifiable data for fields in which I had less experience. I also designed a "test case": If one of the problems is that the behavioral choices needed to address the climate problem in the short term aren't very rewarding from the brain's point of view, can we help people change by making the choices more aligned with what our brains are designed to find rewarding? The resulting "Green Children's Hospital" project was an experiment to see how far this approach might take us, starting with one specific setting. Perhaps the findings also might be applicable to other contexts and scales of endeavor.

Ultimately, the evidence I found pointed to a clear conclusion about the brain: It is a decision-making apparatus that is heavily influenced by its evolutionary design, but is also exquisitely flexible. We are less "hard-wired" than "predisposed," and we spend every moment interacting with and being changed by what we experience. We *can* change what we value and prioritize, and can alter the choices we make, but this is more easily accomplished under some circumstances than others.

The journey that follows—connecting the brain's design to our climate crisis—isn't simple, and readers may find that parts of it are challenging and technical, while others are easier to travel and more entertaining. Through this exploration, I have gained an even greater sense of awe of the brain, along with respect for the idea that we are well served to recognize how the brain's design both constrains and frees us. Understanding these principles may help us change our own behavior, and more effectively encourage others to change theirs, in ways that can help move us forward.

Some years ago, a family sent me a photograph of their child. I had done a risky surgery to disconnect and disable an entire half of his brain, to try to stop seizures that were slowly stealing his language, strength, and intelligence. Thankfully, after the surgery his seizures stopped completely. While he was recuperating, even though he was still weak, he was determined to climb the many steps to the top of a local fire tower. The photo showed the child looking out from the top of the tower over the vast New England forests stretching to the horizon. In the accompanying note, his mother explained that she wanted to share this photo because for her, it captured her son looking out over his new future, one without seizures. But my thought was that we have to give all children a future to look out over—we must pass on to them a planet they can inhabit. These spectacular forests and all they represent won't last without our efforts. That's our job, too. I hope the journey undertaken in *Minding the Climate* helps provide some perspective, some hope, and some ideas for ways forward to address this greatest challenge, one that regardless of whatever else we do, we face together.

MINDING THE CLIMATE

Introduction

The Human Brain and Climate Change

JOHN HOLTER'S SON WAS SICK AGAIN. The boy had been born with myelomeningocele, an opening of the spinal canal to the outside world. This abnormal anatomy disrupts the circulation of the fluid within the brain and spine, so that it builds up inside the head, causing life-threatening pressure. Although the disorder, hydrocephalus, was recognized by ancient Greek physicians, there was still no effective treatment in the 1950s. The artificial tubes used to divert the excess fluid from little Casey Holter's hugely expanded head into his small heart typically drained too little or too much, incited inflammation in adjacent tissues, or simply clogged up, each failure resulting in a new crisis for the little boy.

Casey's father, John, was a precision machinist with the Yale Lock Company. His was a brain on a mission. Long hours spent at his son's bedside, primed by his knowledge of mechanics, led him to notice something about the spongy quality of the intravenous tubing used to give Casey his medicines. This observation led John Holter on a quest to find a material with the precise mechanical properties he suspected would be needed to save his son's life. By luck and persistence, he found what he was looking for—a silicone material called

Silastic that was used in the aerospace industry. Back in his home workshop, with the help of Casey's neurosurgeons, he designed and built a valve and catheter from Silastic that could successfully keep the brain fluid diverted into the heart without the life-threatening complications of all previous treatments for hydrocephalus throughout history. Today, Silastic medical implants are used to treat a multitude of diseases, thanks to the problem-solving skills, tenacity, and drive of John Holter.[1]

From cave paintings to *Beowulf* to classical symphonies to the Mars Rover to decoding the human genome, the human brain has an extraordinary track record of creativity and problem solving. We evolved with unparalleled ability to identify and successfully tackle challenges to our survival, an advantage that underlies our unique ability to inhabit and exponentially populate every corner of the earth.

But our extraordinary brains don't solve all problems equally well. Our evolutionary history equipped us to perceive, prioritize, and find solutions for some kinds of problems more easily than for others, and there are some for which we are ill suited novices. In all our pursuits, we are guided by an extraordinary internal mechanism that evaluates, second to second, our actions in relation to a shifting panoply of human rewards. This complex mechanism, honed by millions of years of history but flexible by nature, assigns value to our choices and guides us with electrochemical currencies that are exquisitely designed under the influence of evolutionary pressure to be fleeting. It is the understanding of that mechanism and how it intersects with our human decisions relevant to climate change that we will pursue in this exploration.

Science, industry, technology, politics, economics, extractavism—climate change is about all these things and many more. It is as wide as the world and as deep as history in its causes and in its scope. But ultimately, climate change is about human behavior. The human brain represents both the cause and the potential solution to this "grand challenge." Scientific details of predictive models can be debated in this unprecedented arena, but the reality of climate change and the preeminent role of human activity are widely accepted around the world. Environmental decline gravely affects society's most vulnerable populations already; it is well along in its inexorable dismantling of ecosystems and populations worldwide.

Our superior ability to solve short-term problems led us to climate change as an unanticipated side effect of our inventiveness; conversely, persistent forms of human behavior remain the major barrier to solving the environmental crisis. If our brains are so able and adaptable, why, then, is it that on average we struggle to acknowledge and respond effectively to an accelerating environmental dismantling, one that comes with critical time limits, and has been recognized as steadily worsening for over half a century? The cause of the problem is not obscure: High-income industrialized countries contribute more than anyone else in the world to global greenhouse gas accumulation and other aspects of environmental decline via ever-increasing consumption. Despite this, our individual and collective behaviors have been slow to change in response to the increasingly urgent consequences of the way we live and the decisions we make individually, institutionally, and politically. Here the premise we will explore is that to understand the paradox of our inactivity in this outward-facing, global-scale problem of climate change, we need to look inward, at how our brains work. Within these insights—about how the human brain perceives and approaches specific types of challenges, particularly those requiring a reevaluation of choices guided by the human reward system—lies potential for change, and some cause for hope.

This book may be of interest to those trying to understand how and why it is that our behavior has led to the current environmental crisis, those who are frustrated that people "just don't change," those who wonder what we can actually do to improve the situation, those who want to know what has been shown to work best to facilitate change, and those who worry that change cannot be accomplished without sacrificing "happiness" as we know it. We will gather evidence from a variety of science and social science fields on the impacts, strategies, and solutions relevant to behavior change and the environment and assess them through the lens of human brain design and function.

Short-Term Changes Built on a Long-Term Brain Design

Technological fixes, including new energy sources and engineered mitigation approaches, clearly are essential to slow the acceleration of greenhouse gas accumulation; these are the focus of intense efforts in research laboratories around the world. Prioritizing and adopting these novel technologies will require overt and large-scale changes in behavior, including revolutions in infrastructure, institutions, and economies. But changing large-scale institutions will take time, and based on the current best climate change predictions, unlimited time is what we don't have before we make essential changes in our behavior.

During the critical decades in the first half of the twenty-first century, before we can overhaul our physical and economic infrastructure so that technology can bail us out, keeping climate change within a range that has a chance of averting the worst catastrophic synergies requires a bridge to decreased consumption that can be adopted quickly, widely, cheaply, and easily. A significant body of research suggests that relatively straightforward measures that already exist to reduce waste, to substitute different behaviors for accomplishing tasks in the residential and workplace spheres, and to consume less in specific categories by those in high income countries may be the most effective—and perhaps the only realistic—way to bridge this carbon output gap. It won't solve the problem, say these scholars, but instituting such measures can result in sufficient climate stabilization to maintain a relatively resilient and recognizable world.[2] Furthermore, these changes don't need to drastically reduce our quality of life. But they do require *change.*

Recognition of the mismatch between the magnitude of the threat of environmental decline and the scale and pace of the response is not new. Scientific evidence for climate change and other intersecting forms of environmental damage has been broadcast at a steadily louder volume for decades, and the percentage of people worldwide who attest that this is a major concern likewise has increased. Accelerating extinctions and loss of biodiversity; burgeoning land use and deforestation; increasing water shortages and ocean acidification; health effects of "forever" chemicals; microplastics in the ocean and on mountaintops; and other metrics of deleterious human influence compete for attention. "Climate depression" and "climate anxiety" now are recognized diagnoses. But the realization that the behavior changes required to address these problems would pose unique challenges has followed a more meandering course.[3]

Climate change in particular plays to our weaknesses. Much has been written about our difficulties in perceiving climate change, due in part to inconsistencies in the information we receive and also stemming from the well-studied tendency to "discount" events that are perceived to occur far in the future or in geographically distant places.[4] These heuristic errors, when we "shortcut" complex decisions involving uncertainty to make them easier to address, have played an important role in our quandary. As we will see, there are additional key features of climate change for which our inherited neural equipment has limited perception. Still, environmental issues are becoming more apparent to more people, as their consequences erupt into daily life for more of the world and as the ever-louder voices of expanding social movements bring them to our attention. But beyond even our challenges with recognizing the problem, our neural engineering helps explain why we have specific difficulties with implementation of the very behaviors that we need to change. Here the case will be made that this occurs because, in short, from the brain's point of view, the behaviors required for this first-of-its-kind problem *just aren't very rewarding.*

A global problem like climate change occurs at a scale and with a complexity that makes it hard to know even where to start, or who should do what, at what level of society. Unparalleled increases in human population constitute an obvious driver of environmental stress. But in terms of cumulative impact, it is the unprecedented increase in per capita consumption from those living in the high-income countries of the world, with the consequent increases in fossil fuel emissions and other waste, that has accelerated the process to critical levels. But to many people, even what exactly is meant by "consumption" in this context and how we go about changing it feels amorphous and unsettled. And for most of us, consuming less doesn't strike us as inherently satisfying.

A strong case can be made that we have faced other big challenges that required major overhauls of the social order and behavioral norms. In the history of the United States, attitude and behavior changes in response to industrialization, racial inequality, and women's rights spread socially in fits and starts, stumbling gradually toward a critical mass supporting a new normal. Global pandemics have spurred dramatic changes in daily life, as well as remarkable pivots in targeted science and technology. Though these problems are difficult and still nowhere near approaching remission, they have features that we are equipped to recognize, and we generally can link responsive individual, moral, and political actions to potential solutions. For climate change action, discounting and other psychological shortcuts slow our progress. But barriers also arise from a discordance between how our brains were designed by evolution to weigh decisions based on survival pressures during a different time in history, compared to the behavior changes required for this unique crisis today. We *can* take action to avert the worst possible outcomes, but the solutions, especially in the time frame required, do not come naturally to us. Our brains may be more readily equipped to make us feel effective and positive by sending money to flood victims than we feel by engaging in the kinds of behavior changes that would help prevent the cause of their suffering. Still there may be ways to pick up the pace if we know why these changes are especially difficult and can implement strategies proven to make them easier.

But change is not purely rational. Why don't people seatbelt their kids? Why don't all motorcyclists wear helmets? Why don't addicts just quit? Why can't we do what we need to do to stop destroying our planet while we still have the chance?

Many fields of study have tried to answer questions about difficult behavior change, from public health, economics, and psychology to government and policy. For climate change, making different decisions with differently weighted priorities is required not just at the level of individuals in their private and work lives but also in their leadership and political roles and as influencers of contagious social movements. It requires change in prioritization for people making decisions as managers of companies, planners in industry, financiers and economists, media influencers, voters, officeholders, and policymakers. But regardless of the scale of influence, the basic unit of behavior change happens person by person.

It's of course true that one person—a persuasive writer, inspirational orator, visual communicator, or institutional leader—may influence many others. Here we focus our attention on the neural mechanisms by which the recipient of new information or new circumstances changes the calculations by which decisions are made, to better understand what elements go into change, and which tend to have the greatest influence, at the individual or group scale. While many researchers have chronicled the reasons people have trouble perceiving the importance of climate change, a smaller number have studied what works best to actually change behavior to facilitate choices with a more direct impact on the climate problem itself. Even fewer have applied a neuroscience lens to understand whether the behavior changes needed may be facilitated by working with, rather than against, the brain's functional design. What is it about how the brain is designed to work that makes this problem difficult for us, and how can we best use that information to help move us in a more effective direction?

The Case for the Brain

Climate change is an enormous, multifaceted problem, and it goes without saying that different disciplines frame the problem and its potential solutions from the perspective of that field. Engineers perceive a design problem, economists see markets and resources, environmental scientists find answers in heat sinks and ice cores, social scientists recognize information flow and group inequities. Each field makes critical contributions to understanding and attempting to find solutions.

Much of the research on behavior change relevant to the environment comes from the field of psychology, with investigators often working in concert with economists and researchers from other disciplines. While classic psychology experiments study behavior observed within a specific time and circumstance, related and overlapping approaches in neuroscience investigate how the nervous system works at the level of cells, molecules, and genes. Psychology describes behavior in specific situations, while neuroscience provides complementary insights into how consistent or malleable behavior may be, based on the plasticity and adaptability inherent in the brain's very design. Whether shifts in behavior are likely to be an effective tool in the climate change battle can be answered only by knowing the environmental impact of specific behaviors, the flexibility of people to make different choices, and the likelihood that enough people might be influenced to change their behavior in a particular direction.

Neuroscience historically has not turned much of its attention to climate change, but the field is steeped in the study of behaviors that are relevant to this problem. Clinicians practicing in neuroscience-based specialties deal routinely with disorders that involve adaptive and abnormal goal-directed behaviors, the influence of experience and neural plasticity on brain function, and other manifestations of the intersection of brain and behavior. Building on painstaking work in basic neuroscience, they treat drives that are "out of balance"—excessive in addictions, dysfunctional after damage to motivation and reward networks—and require strategies for behavior change. As one striking example, patients with Parkinson's disease whose medication doses or deep brain stimulators to control tremors are turned up too high may become compulsive gamblers or shoppers. These disorders shed light on the circuitry and modulation of healthy and "pathologic" brain networks that influence the drive to consume.

In other contexts, clinicians observe daily the amazing resiliency of the human complement of drives and motivations honed also by millions of years of nervous system evolution. Humans are rewarded by agency—the sense of accomplishment afforded by successfully completing a task. Agency almost certainly drove John Holter to take matters into his own hands and find a tangible solution to a problem of exceeding personal importance. But as we will see, perceiving agency from climate change action is a more difficult neural challenge.

Other brain-mediated rewards are similarly critical to the human story. Social rewards are among the most powerful ever identified. And children are especially rewarded when fulfilling their innate drive to explore, learn, and experience. Even after major surgery, what children want most is to go to the playroom and seek out toys; distraction by novelty (most recently, by iPads in the recovery room) has been shown to be more effective than narcotics for reducing pain.[5] There are data demonstrating that exposure to nature is rewarding but that this reward differs in fundamental ways from that of acquisition and consumption. We will explore these neural traits in more detail to learn what factors facilitate different types of behavior change, including those that might have environmental consequences.

This book arose from a particular journey through the topic of environmentally relevant behavior from the perspective of a brain-focused clinician-scientist, specifically in the field of neurosurgery. Not surprisingly, people in this field tend to think *everything* is about the brain—but is that useful? From a brain-centric point of view, behavior-related problems like climate change reflect the design of the equipment we use to interact with and influence the world. Solutions may be enhanced by taking into account an explosion of new insights from neuroscience on how this equipment works at a fundamental level, and how it is or isn't suited to the various responses at hand.

Thus this exploration attempts to address, from a neurobiological point of view, whether insights from the intersection of neuroscience and behavior can help inform strategies to promote individual and collective change during the brief period of time available to alter our global trajectory. From the lens of neuroscience, decisions are arbitrated by the brain's reward system, with inputs from a wide variety of internal and external influences. The brain's decisions and priorities are not predetermined by genes or some unalterable program, and they differ from person to person. Our neural equipment is exquisitely designed to respond to changing conditions—but with certain predispositions and limitations. Understanding the evolutionary design and workings of the brain's reward system in decision making can help us understand the choices humans tend to make in the environmental realm—and most importantly, how malleable these choices may be. While we have some common tendencies to behave in certain ways, we also are engineered to be different from one another by design, as this works best for societal problem solving and survival. In addition, our neural design includes the trait of being highly adaptable to specific types of new circumstances—though there are some strategies that can be called on to make us adapt more easily. Which behaviors contribute the most to environmental harm? What works and doesn't work to change behavior, and why? How fixed or flexible is the decision-making apparatus of the human brain? How does the way we live in current times intersect with our inherited equipment to make things even worse? And finally, if the reward system is an important mediator for behavior affecting climate change, we should be able to create a test case for this hypothesis. Specifically, can we successfully influence decision makers at an institutional level to make pro-environmental behavior more likely, by making it more rewarding?

What If We Don't Change?

Air bubbles trapped in Antarctic ice cores going back 650,000 years contain atmospheric carbon dioxide levels of 180–300 parts per million (ppm). Indirect geochemical measurements from carbon isotopes suggest that atmospheric carbon has not risen much higher than 300 ppm for the past 34 million years.[6] In 2016, daily measurements at the Mauna Loa Observatory in Hawaii exceeded 400 ppm, and CO_2 continues its steady rise that began during the industrial revolution and became even steeper after 1950. The resultant warming already has led to profound weather changes, sea level rise, glacial melting, permafrost thawing, and floods, droughts, and fires that contribute to food shortages, disease, migrations, conflicts, and wars. Enormously expanded human land use due to increasing population and the resultant agricultural and industrial demands decreases the absorption capacity of undisturbed land for CO_2 uptake, decreases biodiversity, upends interdependent natural cycles, impairs species' ability to migrate to escape unfavorable habitats, and creates a feedback loop of ecological degradation. Because released CO_2 remains in the atmosphere for tens of thousands of years, most climate experts plead that unless we find ways to halt the fossil fuel emissions that continue to elevate the earth's temperature at extraordinarily rapid rates beyond what much biologic adaptation is able to withstand, the future is predicted to be very different from all of our human past. More than a million species already have fallen prey to these changes.[7] While conditions head rapidly toward those of the Eocene, the interdependent biologic web is suffering profound disruption, throwing some scholars into doubt about the long-term stability of human societies.[8] We have gotten out of fixes in the past with ingenuity and technology. In pandemics, epidemiologists talk of "flattening the curve"—slowing things down enough so that worst-case scenarios don't overwhelm our capacity to cope. We appear to be caught in a similar time crunch to change our behavior over *decades*, in the hope of giving science and technology, politics, and economics some breathing room to find more long-range solutions, considering the time it will take to get individuals and governments to cooperate and institute large-scale changes. But even on smaller scales, behavior change is hard, and we need all the insights we can bring to the problem.

This book, then, is the journey in search of answers to these questions. As this linkage between neurobiology and climate change is a relatively untrodden path, we won't find a neat answer to every question. We won't be able to plumb the depths of every field—basic neuroscience, evolutionary biology, animal and human psychology, behavioral economics, child development, sociology, consumer behavior, environmental science, and others whose scientific advances we pass along the way. But we can use some pertinent findings from each field as examples of general principles that can inform our progress. The relation between fields won't be uniform and linear, and our path will circle back to ground we've already covered from time to time, to find the linkages between these bodies of knowledge. We will encompass in our exploration the parallel journey of human evolution and the serial refinements of the workings of the human brain, which will shed light on how and why we got here, and how we may be able to use this information to help change our course.

We embark on this path to investigate whether it offers anything of value to help us more effectively meet this vexing crisis of our time. As our first step, we will next start our story together at the earth's inception.

Part 1

Neural Origins

1

Brain Evolution and the Anthropocene

OUR PLANET IS THOUGHT TO BE ABOUT 4.5 BILLION YEARS OLD, while anatomically modern humans first appeared about 200,000 years ago.[1] The earth has been in existence over 30,000 times longer than humans have. Although we are relatively recent residents of our world, we are designed and built on an evolutionary framework that predated our species by billions of years, shaped by biological forces far different from the challenges that we face in the present. We are living now in a period of earth history which has been called the Anthropocene—the geological era of human-induced changes to the physical world and its atmosphere.[2] Yet our brains evolved to overcome very different challenges than what we face today. While climate change may seem like a slow and gradual process compared to the timeframe of our own lifetimes, it has happened extraordinarily rapidly when stacked against the very long timeline of the planet and the life forms on it, including humans. Thus, our brain design, shaped during its long history by different evolutionary pressures for survival, has not had time to evolve to meet the very recent and unprecedented challenge of climate change, the most globally impactful aspect of the human-induced Anthropocene. Understanding why and how our brains work as they do may help us home in on the most effective strategies to change our behavior to meet such a dangerous and novel threat.

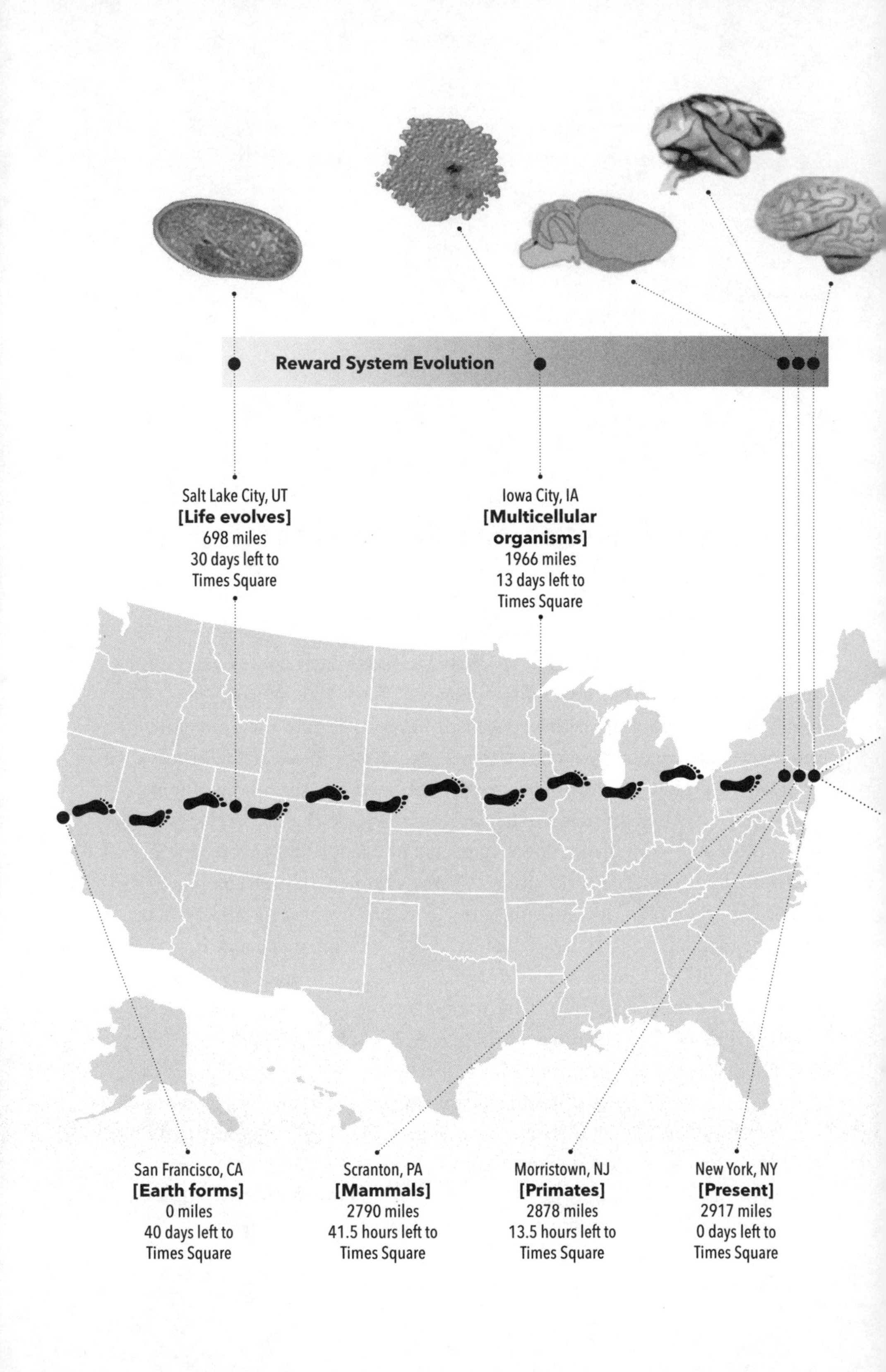
Reward System Evolution
Salt Lake City, UT
[Life evolves]
698 miles
30 days left to
Times Square
Iowa City, IA
[Multicellular
organisms]
1966 miles
13 days left to
Times Square
San Francisco, CA
[Earth forms]
0 miles
40 days left to
Times Square
Scranton, PA
[Mammals]
2790 miles
41.5 hours left to
Times Square
Morristown, NJ
[Primates]
2878 miles
13.5 hours left to
Times Square
New York, NY
[Present]
2917 miles
0 days left to
Times Square

Fig. 1 Brain evolution timeline compared to the Anthropocene, the era of human-caused climate change and other anthropogenic alterations. The history of the earth from its inception to the current time is pictured as a forty-day walk from San Francisco (the earth has just formed) to the center of Times Square (now). The evolution of different forms of life with accompanying brain evolution is shown on the timeline and at the top of the figure. The Anthropocene begins when the big toe hits the ground during the last step, or the last 0.18 seconds, of the forty-day walk. Corresponding changes in atmospheric carbon appear below the expanded timelines shown in the far-right insets. Atmospheric carbon has continued to increase sharply since these graphs were made, reaching 419 ppm in February 2022 at the Mauna Loa Observatory in Hawaii.

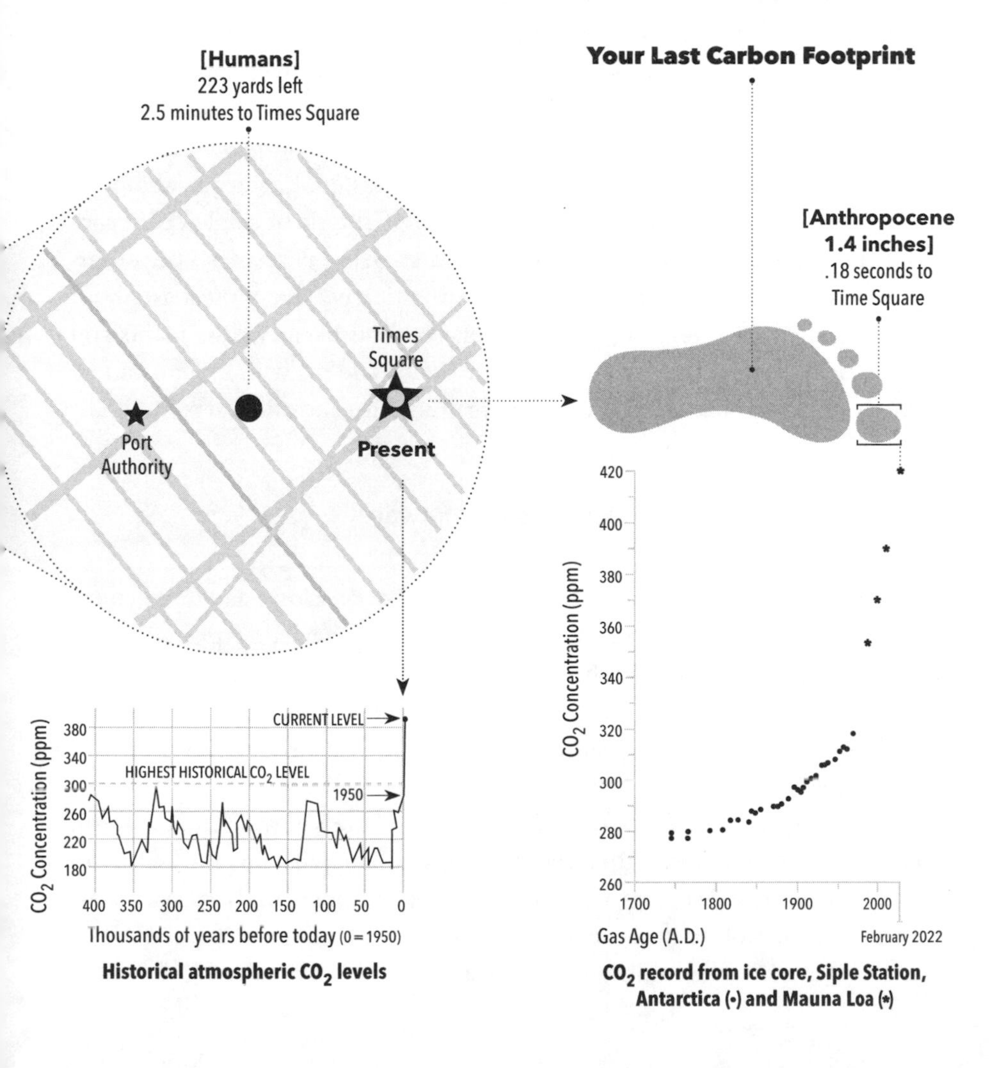

To get a sense of the relative times involved in the evolution of our brain compared to the Anthropocene, imagine walking from San Francisco to New York at a steady pace (Figure 1). The trip takes forty days, nonstop. If the starting point in San Francisco represents the origin of the earth, and the center of Times Square in New York City is the present, life on earth started at the time you reach Salt Lake City. Multicellular organisms, including those in which the underpinnings of our modern reward system first appeared, come along in Iowa City, a little after the halfway point of your journey. Mammals appear in Scranton, Pennsylvania, within two days of the end of your walk. Primates make their appearance in Morristown, New Jersey, on the last day of your trek, and modern humans just after 42nd Street, with 2 ½ minutes left to go. In contrast, the Anthropocene occurs only in the last 0.18 seconds of your transcontinental walk—when your big toe hits the last part of its final step.

The evolutionary process that shaped our brains began long before humans appeared; much of the way our nervous systems work today was inherited and refined nearly as long as living organisms have existed. Life began about 3.5 billion years ago with the formation of the simplest type of single-cell organisms, prokaryotes, followed about 2 billion years ago by eukaryotes—organisms that incorporated a membrane-enclosed instruction packet, or nucleus. About a billion and a half years ago, multicellular organisms, including plants and animals, appeared. The evolution of our brains and nervous system can be traced all the way back to single-celled organisms like bacteria and paramecia.[3]

Ancestors and Reward

The way in which our brains work to think, make decisions, and take actions may seem mysterious and magical, but there are mechanistic explanations for how these tasks are accomplished. To gain insight into our predispositions and our limitations, we can start by observing how the complex machinery for assessing choices was assembled by evolution through the history of living organisms.

Strategies to assess the outer and internal conditions in which an organism finds itself, and the means to make "decisions" about behaviors most likely to lead to survival and success, occur in the simplest organisms. Bacteria, for example, which evolved near the half-way point in our timeline billions of years ago, can move toward nutrients and away from toxins. Understanding how things work in simple systems is crucial for understanding how these basic building blocks work to help direct our choices in humans today. How does a single cell "know" what's around it, and "decide" whether it stays still, moves forward, or backs up? It does so by using sensors, which are typically molecules somewhere on the organism's surface that have the property of changing shape when they encounter specific chemicals. So much of how we and other living beings function arose from fortuitous combinations of molecules with specific properties that are altered by changes in our surroundings. These principles form the basis of the amazing interactions that underpin all biological processes. In bacteria, this change in shape of a sensor molecule on its surface causes a chain reaction with other interconnected molecules such that tiny "propellers" on the cell body move the cell toward the direction of the sensors that encountered something attractive to the animal, like sugar. Different sensors evolved the ability to detect something harmful, like toxins, and then the chain reaction causes the propellers to reverse direction. Early life forms that happened, by chance, to acquire a mutation that equipped them with this kind of molecular sensor and also had a way to act on the information it provided survived better than those without, so the genes that coded for this advantage were passed down, and gradually improved upon, following the principles of natural selection. Thanks to intricate techniques that allow scientists to determine how minuscule proteins change their three-dimensional shape when chemical reactions occur, we now understand the choreography of the entire set of reactions that make all these options possible in an organism that has no more than one cell, never mind a brain. In even more amazing complexity for a single cell, these simplest of organisms evolved to choose survival over a good meal, and if both nutrient and toxin sensors are activated at the same time, the chain reaction is designed to skedaddle.[4]

Over time, these same principles seen in single cells were refined in larger multicellular organisms with more body parts, more behaviors in their repertoire, and a wider range of environmental conditions in which to navigate. These creatures had more possibilities to experience, more things to encounter, more decisions to make, and more learning to do. As the tools developed to study them, scientists were able to work out the entire nervous system and the role of each of the 302 nerve cells of tiny creatures like the nematode *Caenorhabditis elegans* (*C. elegans*).[5] This entire beast is 1 millimeter long (about the width of the number "1" on this page) and lives its life moving about in the dirt. The decisions made by animals of this scale include whether to move forward toward something "good" (often, tasty!) or back up away from something "bad" (an obstacle, a dangerous chemical, or a predator), whether to move in a different direction to explore or to stay still and feed, and how and when to reproduce. How do 302 neurons accomplish this? In short, they are specialized to sense and provide information on important aspects of the outside world, and they communicate that information to a nerve cell network that is able to evaluate the "relative value" of the various options, taking into account internally generated signals that are sensitive to fluctuating conditions within the multifaceted state of the animal. This nerve cell network then has the ability to communicate the "optimal choice" to specialized nerve cells that enable movement or another desired behavior to occur. To accomplish this, the nematode uses chemicals that act as neuromodulators—substances that change in subtle ways how the nerve cells work. In the nematode and many other animals, these are octopamine and dopamine, chemicals that are also found in the human brain, doing much the same kind of thing on a larger scale.

A substance that acts as a neuromodulator can alter how easily signals received or sent by nerve cells are perceived, transmitted, or act. Similar to a booster or lubricant influencing how an engine performs, neuromodulators are specific types of chemicals released from the ends of neurons that act on specific other neuron connections to make them work together better—or to slow things down, depending on the specific situation. Neuromodulators are critical to how the reward system works, both in simple organisms and in humans.

In a neurologic context, all behaviors, such as the coordinated set of movements involved in eating or moving in a specific way, occur as the direct result of a series of neurons activating, or "firing," in a specific sequence. This is best visualized as a complex "net" in which many things are happening down different interlinked pathways nearly simultaneously. As we will explore in more detail in Chapter 2, neurons are individual cells specialized to communicate with many other neurons through small gaps (synapses), across which impulses can pass. These networks are spread throughout the nervous system; a firing pattern associated with a specific behavior may use one set of "highways" and another behavior may use some of the same and some different neuronal paths. Specific neurons can cause muscles to contract, substances to be released, and signals to be sent to other parts of the body. Specific patterns of neurons firing in sequence can cause complex thoughts and behaviors; picture thousands of tiny moving lights zipping along a complicated network of interconnecting nerve cells.

Often neuronal signals feed back to where they came from in some way, forming an electrical loop, or "circuit." While not every functional part of the nervous system necessarily uses a circuit design, these feedback loops are a common feature of many parts of the brain and its connections into other parts of the body, including motor and sensation circuitry, brain hormonal systems, and the reward system (sometimes referred to as the reward "circuit"). The brain uses circuits in order to respond second-to-second to changes in the environment, to fine-tune its actions, and to adapt to internal and external signals. Circuits make you hop back the instant you step on a Lego. If it weren't for this kind of circuit, which, among other functions, helps dampen your perception of things that aren't changing or important, you'd continuously feel every tag and fold in your clothing, all day long.

Much of the basics of reward biology were worked out in learned feeding behavior in simple organisms like the foot-long giant sea slug, *Aplysia,* because its large nerve cells and relatively simple circuits can be studied one neuron at a time to understand the sequence of events. If something you do leads to eating something nutritious, it would be helpful for you to repeat that behavior. Learning happens because circuits that are activated by chemical interactions with the outside world, and linked with something "good" happening—such as swallowing something nutritious—start a chain reaction by which the neuron sequences—the behaviors that happened just before that "good" thing happened—are strengthened.[6] If the circuit can be imagined as a big set of roads, and the firing as the path of a car, neuromodulators smooth the intersections so that particular patterns happen more easily the next time—the car has an easier time traveling down a particular path among all the roads available. This, then, is the beginning of a reward system—a set of neurons that gathers inputs from various aspects of the outside and internal world, and are designed to process this information, weighing it to make decisions about how to value and act on the possible options. The system is designed for two main tasks—first, identifying and reinforcing, or "rewarding," a beneficial behavior so it is more likely to be repeated. Second, the system is designed for learning about what is valuable and important in the outside world. Through minuscule chemical changes in the connections between neurons in different parts of the nervous system control center, modulation by the neurons in the reward system increases the likelihood that a behavior that leads to something beneficial for survival in the short term will be repeated. The modulator chemicals released by the reward system neurons also cause connectivity changes that strengthen associations between that behavior and the overall sensory signature of the circumstances under which the "good" things occurred. A classic axiom to describe this is "neurons that fire together, wire together."[7] When events happen in just the right sequence and circumstances so that the reward system perceives a pattern as resulting in something "good," the chances that a specific linked pattern like eating will be tried the next time those circumstances are encountered are increased. This is why you are inclined to return to a restaurant at which you had a particularly satisfying meal, to a social setting where you met a charming person, or to the home of someone who complimented your child. Not all of this necessarily occurs at a conscious level; you may not even remember the compliment, but when you think of the person and her home, you get a "good feeling." Your reward system has "taught" you a positive association that can be triggered by the memory of that person, or even elicited by the sound of her name. We will learn more of the details of how this works in humans, and how it can be altered, in the next chapters.

As our nervous systems were being shaped during prehistory, what information was sensed, and what decisions were made and which future behaviors were made more likely, were all fine-tuned in line with evolutionary principles over millions of years. This means that time favored those nervous systems that led to choices with the greatest correlation with survival and reproduction. By taking advantage of random mutations caused by genetic variability, those individual organisms with better or more accurate sensors, better reactions, or better decision-making formulas had the advantage in survival, and so their design had an increased chance of being passed down to subsequent generations of offspring. Thus reward circuitry, like all the other aspects of brain and body, was passed on ("conserved") and gradually became specialized for specific niches as different types of animals evolved to meet distinct competitive challenges. For this reason, a great deal of commonality among animals exists today in the mechanisms for recognizing, at the sensory and brain processing level, something as "good" or "bad"—as rewarding or noxious. Speck-sized fruit flies and thimble-sized crayfish can learn associations between their own actions and specific rewards like food. They also can learn associations between a neutral stimulus, like a sound or a location, and a reward, like food, such that the sound or the place itself becomes rewarding once it has been paired repeatedly with food. And the neural mechanisms they use to do this echo what exists in human beings.[8]

But aren't there things that are "naturally" rewarding? Food is the classic example of a so-called "primary" reward—when an animal gets food, it was long thought that it doesn't need to learn that food is rewarding—it just naturally is (as long as the animal is hungry). But from the neuroscience point of view, nothing just happens—there are clever ways nature accomplishes these tasks. Selection in specific evolutionary niches might make one type of food more rewarding to one type of animal compared to another—plankton, eucalyptus leaves, carrion, honey. Despite these differences in primary reward, all healthy animals have the capacity to learn—and so, for each creature, some things become rewarding because of their learned association with a primary reward; this is called a "secondary" reward. This allows the animal to learn under what specific circumstances those things necessary for survival might be found, as these may vary for individual animals or may change over time and require learning new associations. Even actions as basic as eating and drinking are learned in newborn animals—but evolution made the pairing of these behaviors with appropriate sources in the newborn so typical that the pairing happens almost always, making it more difficult to detect the "learning" component.[9] It is the reward system's task to make sure these associations happen.

Simple organisms and more complex organisms, including humans, have reward systems with an important characteristic in common: they are designed for *short-term decision making*. Something happens, the system evaluates the various inputs, adjudicates conflict—and the system responds. Associations are learned, and remembered, and if the pairing with reward ceases, most of the associations are unlearned over time. If, during the evolutionary history of an organism, circumstances changed—for instance, if a particular food source were to become scarce—individuals who could metabolize an alternate food would survive, if the reward system also evolved to "stamp" that food source as "rewarding." Otherwise, they would waste all their energy looking for something too scarce to provide for their needs. Similarly, if water became more acidic, those best adapted for survival would be those who were more acid tolerant and who also could learn that a lower pH was not toxic—otherwise, they'd spend too much energy moving away from something that wasn't really a bogeyman after all.

How did these changes occur during evolution? To bring this discussion back to our overall timeline, the time scale over which evolutionary adaptations occur is very, very long. Mutations, the tiny spontaneous changes in the instructions carried in the DNA, occur quickly—in fact, all the time, in each new individual organism, including in each of us. But the shaping of which mutations in the DNA instruction manual are actually expressed as a new physical change—and are preserved and passed on to offspring to create a lasting evolutionary change—is a very slow process, typically over hundreds and thousands of years. To make the cut, a mutation has to provide an advantage to that animal under those specific circumstances, occur in sync with other mutations that are also needed to make everything work together in a coordinated manner, and confer the advantage during the time that the animal also happens to reproduce. A classic example is the Galapagos finches that had thicker beaks and so could eat the harder seeds produced during drought. When a drought happened, they survived more often than the finches with thinner, finer beaks, and so the characteristics in the population shifted.[10] So while there are countless mutations, most don't last, and they die out with the individual whose genetic material happened to undergo that tiny change. Only a few alter the animal in a way that ends up causing a visible change that confers an advantage and thus perseveres.

Mammals carried on many of the basic designs of the reward system present in their distant ancestors. On our transcontinental walk, mammals show up in the state of Pennsylvania, about thirty-eight days into the forty-day journey. As the space and variety of circumstances in which an animal navigates expands, so does the variety of experiences, choices, and range of things to find rewarding or noxious—in other words, so expands the scale of necessary learning. For example, the rat has an enormous repertoire of behaviors and choices in its daily life. Where and when and how to forage successfully for food, what to eat and what to avoid, how to get from food to safety to home, how to respond to challenging weather, how to recognize and avoid predators, how to socialize, when and how to select a mate, how to raise young, when to nurture and when to be aggressive, how to make decisions when conflicts between potential behaviors arise—the list is almost endless. Nervous systems in more complex animals with more complex lives have to master not just conditioned behavior—this external stimulus is associated with that reward—but also so-called goal-directed behavior. This requires learning that a specific action results in a reward, and thus behavior becomes directed at attaining the reward. Finally, making decisions and choices among possible behaviors requires the capacity to make judgments about their potential relative value, or utility, to the organism at a given point in time.[11] As new species evolved and spread, and their spectrum of behaviors became more complex, the reward system also expanded to facilitate the learning and adaptation, judgments, and choices most relevant for survival during natural selection.

It is important to recall this basic principle—that the reward system evolved not to make life pleasant or "rewarding" in the general use of that word, but as a means to make a behavior that increases the individual organism's chances of survival more likely to occur.

Climate change—the last fraction of a second on our forty-day timeline of our planet—is "fast" with respect to the history of the earth, but "slow" with respect to how it affects our day-to-day, immediate survival. Thus it should not be surprising that the components of the human reward system have not been designed by evolution to perceive, process, or act effectively on a relatively "slow" and gradual change in immediate survival circumstances, such as a problem like climate change. This is not just about our tendency to discount future events; we will discuss in the next chapters why unfamiliar events present an even greater challenge. But an even more basic factor is simply that the cause-and-effect linkage between environmental threats and the behavior needed to respond is neither direct nor obvious. Throw a rock at a predator, which then runs away? This makes sense. Your reward system is set up to learn that throwing the rock worked, and you are likely to repeat the behavior next time you are in the same circumstance. Bike to work to prevent global warming and food shortages? The efficacy of this behavior is much more indirect and difficult for us to perceive and learn, to the point of seeming silly and even self-righteous. It's not just that the events are in the future (which increasingly is not the case anyway)—it's that the behavior and the consequence are difficult to tie directly to the experience of the behavior in many dimensions—temporally, spatially, conceptually—even motorically. It is this kind of mismatch that is in part at the root of why taking appropriate action has been particularly challenging.

In the next stops on our journey, we examine how the mammalian, and specifically, human reward system works in more detail. This will enable us to look at how behaviors relevant to consumption and environmental decline in its various forms may be influenced by our evolutionary, social, cultural, and individual histories. We will then be in a better position to evaluate if and how those things we find rewarding can be modulated in ways that are relevant to our environmental quagmire.

2

Brain Rewards as a Design for Learning

HUMAN ADULT BRAINS CONTAIN ABOUT 86 BILLION NEURONS, with unfathomably enormous numbers of connections—each neuron having multitudes of processes which can connect to thousands of other processes on other neurons.[1] To maximize adaptability, children have an even greater number, and prune during development to select those brain circuits most useful for their specific life circumstances—hunting with a slingshot, learning two languages, playing the violin. In comparison, the brains of mice and rats, from which much of our understanding of neural structure and function is derived, have 70 and 200 million neurons, respectively. It is the enormous complexity and plasticity of the human brain, which reached its current structure in the last few minutes of our forty-day walk, that provides the basis for our unique abilities and for the variability of one person compared to another. This plasticity underlies changes in an individual person over time, which is the key ingredient to changing behavior—even the behavior of whole societies.

What do the structure and function of the brain have to do with a monumental societal problem like climate change? The brain's plasticity, and the reward system that drives our behavioral choices, have everything to do with changing behavior. It is the degree and speed with which we can change human behavior, at the individual, political, and collective scale, that will determine how we face this crisis.

It's obvious that despite all having the same "model" of brain, people act differently from each other. The differences in environmental metrics between Europe and the United States do not reflect an inherent difference in brain equipment across the Atlantic, but instead reflect social, economic, political, and informational differences—all of which are filtered through the brain to result in behavioral choices. Your brain's underlying temperament and talents are different from mine, and our brains' adaptations to life experience differ, but we both have all the same parts and very similar blueprints, and our cells work with certain common predispositions. From the perspective of neuroscience, we might say this about the brain and social problems: Understanding the basic principles of how we make decisions, what predispositions we carry, what genetic and environmental factors modulate behavior, how we learn, and how we react socially and emotionally to new information and to the influence of others, are all parts of the puzzle of understanding problems and evaluating the potential efficacy of possible solutions, including those that occur in the social realm. Understanding how reward system function influences those 86 billion neurons is just another way to work toward understanding people.

To get a sense of how this system works, how it evolved, and how it can be changed, we will start with some basic ingredients and then explore how they work together to determine our choices and behavior. These mechanisms are utilized in all our decisions. As we explore them, we can imagine that we are looking into a specific brain—for instance, your own brain as you make a purchasing decision, that of a company executive deciding whether a more environmentally sound operational process is too expensive, or that of a politician about to vote on a bill enacting regulations that will help curb climate change. Each behavior is preceded by millions of lightning-fast steps that reflect your evolutionary and personal past, your current circumstances and physiological state, your genetic predispositions, and factors of which you're aware and can reflect on consciously. Most decisions involve making predictions involving uncertainty—in essence, making an informed gamble. Our politician knows that in the short term, the measure will increase prices on certain commodities, may cost some jobs, and likely will incur the wrath of some local businesses. In the long term, maybe some alternative jobs will be created, maybe health will improve, and there may be fewer extreme weather events, about which some voters in the district have expressed concern. Subsequent chapters will address consumption, climate-relevant behaviors, and behavior change in broader and less technical terms. In this chapter, in order to understand the basic mechanisms by which we make decisions, we will explore some details about how and why these processes work as they do. Let's go deep into the brain to watch what's happening as it faces these choices.

Brain Ingredients

Neurons, the main "action cells" of the brain, come in many shapes, sizes, and varieties. Think of all the different kinds of nails in your house—there are huge ones, squat ones, tiny ones, with different shaped heads and points. Similarly, neurons can be specialized to respond to different types of sensory inputs, to transmit specific types of messages at particular speeds, to process or store information, to modify how other parts of the system work, or to interact with other parts of the body such as muscles, blood vessels, or glands. While neurons are highly specialized, most have three main parts. These include a cell body that serves as the main instruction and manufacturing center, processes called dendrites that act like antennae to collect information from other neurons, and an axon process that sends signals to other neurons or targets. Collections of axons have insulation (myelin) to help signals go faster and are called "white matter," because the fat-based covering looks whitish when a fresh brain is cut in slices to look inside. The neuron cell bodies with their associated complement of energy-providing blood vessels look pinkish gray in color, so layers or collections of neurons are called "gray matter."

In most mammalian brains, including that of our politician on the verge of casting a vote, the entire surface of the brain closest to the skull in all directions, the cerebral cortex, is made up of multiple layers of neuron cell bodies, along with astrocytes and other types of cells that facilitate brain function. If neurons were people, the cortex is billions of them standing side by side, with their cell body "heads" floating near the inner surface of the skull, their dendrites like arms reaching out sideways to touch other people, and their legs sweeping together with billions of other axons to form the white matter tracts that run toward the center of the head and through the bottom of the skull to connect with other parts of the brain and body. Other neurons are designed differently for specialized roles; interneurons, for instance, live between other neurons and modulate their function.

In humans, the cortex is divided into frontal, temporal, parietal, and occipital lobes. While highly interconnected and with multiple integrated functions,

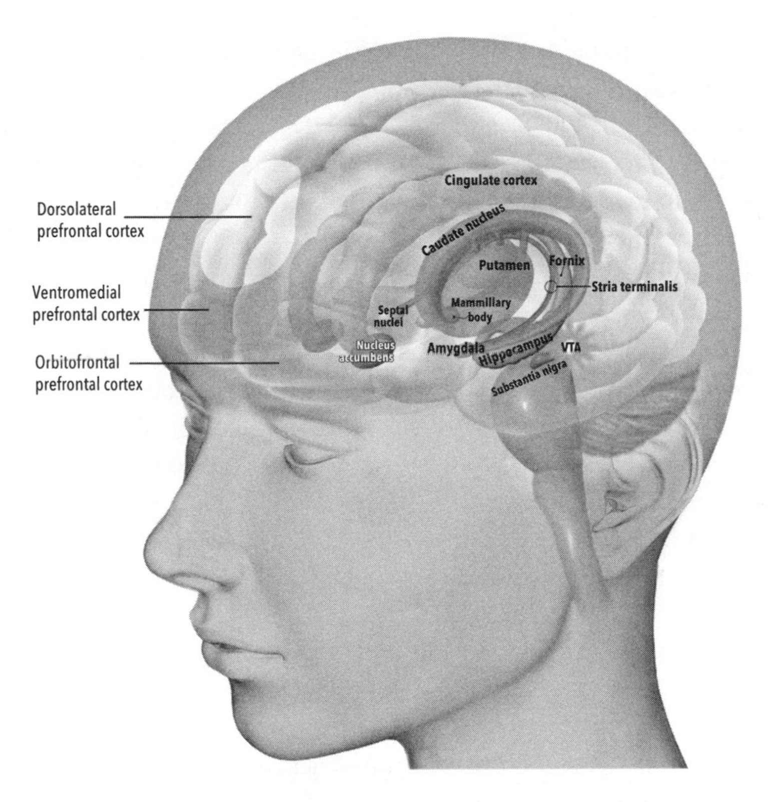

Fig. 2 Anatomy of the human reward system. Details can be found in the text.

in general, the frontal lobes subserve motor and many cognitive functions including decision making; the temporal lobes are involved with language and memory; the parietal lobes are concerned with sensation, spatial perception, and many kinds of cognitive processing; and the occipital lobes with vision. The cortical lobes are further subdivided into different areas, as shown for the frontal lobes in Figure 2. We will see how the intricate connections between regions enable their specific functions, such as deciding among choices.

The cortex and other collections of gray and white matter that we will be discussing live in the large, dome-shaped upper chamber inside the skull, surrounded by protective fluid-containing membranes. These structures are partly separated from those in the lower part of the brain by a tough, tent-shaped membrane. Below this tentorium lives the cerebellum, a cauliflower-like structure important for coordination, balance, and motor learning, among other functions. Also in this smaller, lower compartment is the *brainstem,* through which course essentially all the functional to-and-fro connections between the brain and the body. The brainstem also contains relay centers for nerves in charge of eye, face, and tongue movements; breathing, swallowing, heart rate, and alertness; and some vital components of the reward circuit, as we will see in more detail shortly.

In biology, some words have more than one meaning. Within an individual cell, the *nucleus* is the membrane-enclosed packet of genetic instructions. But when describing the gross structure of the nervous system, a nucleus is a collection of cell bodies, typically deep in the center of the brain or brainstem. Some larger structures made up of collections of nuclei get their own names, such as the walnut-sized thalamus with its multiple thalamic nuclei. By the historical quirks of how structures get their names, other similar collections of cell bodies may be called by a different name, ganglia, such as the basal ganglia, important in reward circuitry, as we will see.

How Cells Converse

The action in the brain occurs via communication between all the different parts. Its complexity and extraordinary design features are what make decisions like the one our politician is facing possible—and, as we will see in the next chapters, not always "logical." The communication point between neurons, the synapse, includes a specialized portion of the axon of one neuron, a specialized portion of a dendrite of another neuron, and the salt-watery fluid between them. Most synapses don't work by direct electrical connections, like a spark jumping a gap. Rather, they are chemical connections, and work by the release of neurotransmitter chemicals from one neuron that interact with specialized receptor sites on the next neuron. Multiple types of neurotransmitters exist, each with different properties. On the receiving side of the synapse, different kinds of receptors, like specially shaped locks, wait to interact with a specifically shaped key—the specific neurotransmitter to which it is designed to respond. Earlier we met the neurotransmitter dopamine, critical in reward circuitry.

The different types of receptors provide specialized responses to the neurotransmitter released from the "upstream" (or presynaptic) neuron in the synapse. Most of the receptors, when bound by their appropriate transmitter, change shape to let something that is otherwise blocked enter the "downstream" (postsynaptic) cell for a short time, like a gatekeeper opening a gate. We saw this same basic principle at work in single-celled organisms in Chapter 1, where receptors on the cell surface interacted with substances in their environment, changing shape to let in specific molecules needed for the cell to move in a specific direction. Similarly, in synapses, often the things allowed through the gate are specific charged ions, such as sodium or calcium. When enough receptors on the postsynaptic neuron are similarly affected, its electrical charge can reach a threshold such that the cell propagates an electrical impulse that spreads throughout its whole cell body and axon. This action potential, or "firing," in turn causes release of neurotransmitters from the axon at the end of that neuron, and this can affect the other neurons with which that cell has synaptic connections. This overall form of communication is often referred to as "electrochemical" transmission, because it uses both chemical and electrical processes. But, you might think, this is so complicated! Why does it have so many parts and pieces and options? As it turns out, this is an extraordinarily well-engineered strategy to balance energy efficiency in the brain with rapid speed when needed, so that you can find enough calories to sustain its enormous computational power and still respond fast enough when a speck of dirt is blown toward your eyeball.[2]

But here's the reason your brain is so complex and can perform such high-level tasks and adapt to so many circumstances. Even when the second cell in line fires, it doesn't necessarily mean that there now will be a chain reaction with multiple connected cells firing—and a decision will be made, just like that. Each neuron has input from multiple cells, and the release of neurotransmitter from any specific cell only tweaks the next cell, and the whole system, a very tiny amount. It has been estimated that each of the 86 billion neurons makes about 10,000 synaptic connections with other neurons. It is worth pausing for a minute and processing this fact—86 billion neurons, each with 10,000 connections. That means a given neuron will be getting input at any given time from thousands and thousands of other neurons. Neuronal firing is no knee-jerk reaction. Every cell, every moment, every thought, and every decision depends on so many minuscule events that to date, the outcome of these complex processes are well beyond our human capacity to predict.[3] We can study behavior and make educated guesses based on probability—that most rats or most people in a given constrained set of circumstances will do "a" or "b." But when we say there is variability, this reflects the enormous complexity of inputs, connections, and outputs, which vary widely among individuals and over time.

This very complexity predicts that there will be variation from one person to the next, from one moment to the next, and that there are reasons for the amazing range of behavior that humans exhibit. We also can understand why certain types of behavior tend to occur along certain lines, particularly, under specific social and cultural influences. While there is tremendous plasticity and modulation, these influences occur in the context of specific inherited biological predispositions. When it comes to reward and consumption, there are principles based on survival value that we all inherited—and these predispositions help explain how human decisions and behaviors have led to climate change. But how we will actualize those tendencies going forward will vary enormously based on the incredible breadth of circumstantial inputs on which behaviors depend.

Within the range of our inherited predispositions, everything can change, and does so, all the time. Both the axon releasing the little batches of neurotransmitter and the receptors in the next neuron in line can constantly change. The synapse is a living organism, and is rebuilt and modified continuously, like a toll booth constantly under construction. This change is under the control of the instruction manual of the cell, typically its genetic material and other parts of the cell machinery, which also are not fully fixed but instead constantly respond to external and internal cues, changing in response to all kinds of events. What you ate—in fact, what your grandmother ate—can change how your genetic book of instructions is translated into the proteins that determine how the cell works and how it responds. Receptors for different neurotransmitters or with different properties (for instance, how quickly or for how long they "open the gate" when attached to a transmitter) can be churned out at different rates in response to all kinds of inputs from the outside or inside world that finely adjust the cell body's machinery. These kinds of changes are responsible for why you stop paying attention to a new background noise after a few minutes, why there may be changes in behavioral tendencies lasting for generations after a famine, and why some things that you experienced this week will stick in your memory for years.

Why can't you always convince people to make a change in behavior by simply explaining to them why they should? This can create frustration for advocates of particular issues, including pressing and urgent issues like climate change, when compelling arguments are put forth backed with facts galore and impeccable logic. But the complexity of the human decision-making apparatus makes any one influence only one part of the equation. All of the overt and opaque factors noted above will play into the decision that our legislator will make momentarily. To understand how all the millions of events are weighted, we next turn to the mechanism that evolution provided us for learning what we need to know to make decisions that will help us survive—the human reward system.

Thinking in Patterns

Have you ever had the experience when you're walking down the street and see someone, and for a brief instant think it's someone you know? This can happen even when it's impossible—say, someone who just moved to China or it's your grandmother who died years ago. But for that brief second, your brain says, "Hey—it's . . . !" until you realize it can't be. Something similar happens when you're walking along and think for a second there's a dog down the street, before you recognize it's an overturned trash barrel. Or when you find yourself internally humming a song you haven't thought about in years, for no apparent reason, but if you retraced your steps and listened carefully, you'd find two workers hammering at a construction site happening to tap out the first two notes. You weren't aware of any of this consciously, but your brain took in these inputs, and ran with them.

All of this happens because your brain thinks in patterns. Those astonishing numbers of neurons and synapses strengthen with experience and predisposition, creating lightning-fast networks of neuronal connectivity that can be triggered by any number of sensory inputs that "match" some part of a pattern you have stored in your brain. If your grandmother was wonderful, that triggered pattern also will loop in a positive emotion, and your face will light up for that millisecond in which you think you see her down the street. If the song was associated with a good time in your life, your mood will be lifted for reasons you couldn't even describe. That we recognize patterns everywhere, second to second, and filter out those that don't make sense or match what we know (that wasn't really a dog) reflects that this trait—matching what we experience to what we've experienced before—is extremely useful for learning and survival. The reward system plays a critical role in establishing, maintaining, and changing these patterns.

The History of Scientific Progress: Deciphering the Human Reward System

Scientists have pieced together how the reward system works to help us learn and make decisions from several contexts. This amazing story has unfolded as researchers discovered new clues about the structure and function of how humans assess choices—including those relevant to our planetary future. Observation of patients with conditions that affect reward-related behavior, research using simple and more complex animals, and addiction research that investigates reward function gone awry all provide insights. Different tools analyzing patients with specific diseases, animal models, brain imaging, and recording from single cells and networks each has shed light on the detailed neurobiologic processes that underlie how we make decisions and what is and isn't easy for us to change. These biologic processes manifest in behavioral choices, and researchers in economics, marketing, advertising, public health, and policy investigate what external factors influence people's choices so that they can be changed in a desired direction. Finally, recent decades have produced a body of research specifically on the concept of happiness—is there a difference between short-term "reward" and more long-term life satisfaction?

While entire libraries can be filled with detailed research on reward and decision making, we will briefly visit some of the approaches that have provided a map of how our brain reward system works, and for what purpose. We will highlight discoveries as they were made roughly in chronological order, as successive pieces of the puzzle were understood using new scientific tools and building on what came before. Having a sense from these various fields and approaches of how the system works will equip us to assess the alignment between how this system is designed to function, and the choices and behaviors most likely to affect our global environment. Can the human brain change what it thinks is most important, and if so, how does that happen?

Early Clues from Human Patients

For thousands of years, physicians have known that injury to the head could cause problems with consciousness, strength, and sensation. However, the relationship between particular parts of the brain and more psychological functions (such as affect, behavioral regulation, judgment, and decision making) were poorly understood until relatively recently. The frontal lobes were implicated as important in personality and judgment by famous cases like that of Phineas Gage, an industrious 25-year-old railroad worker who in 1848 survived, remarkably, a through-and-through transection of his left frontal lobe by a 3-foot-long gunpowder-driven tamping iron. The pointed iron rod entered his cheek just below his eye and exited through the top of his head; he sustained only a momentary loss of consciousness and was awake and talking at the scene. Miraculously, the course of the projectile missed the major blood vessels and vital structures that would have led to immediate fatality. Beyond that rare bit of luck, Phineas survived because his progressive doctor took a gamble and surgically drained the brain abscess left behind by the dirty tamping iron, providing the patient's only hope in the pre-antibiotic era.

After recovering, Mr. Gage exhibited some personality and behavioral changes that remained the source of acrimonious debate in the medical world for years, reflecting competing theories about whether specific functions are localized to specific parts of the cerebral cortex, or whether instead cortical regions are interchangeable.[4] This argument demonstrates a principle in science, including medicine: when people argue vehemently, it usually means there are gaps in knowledge that preclude a full understanding of the "true" answer. What has been learned since Phineas Gage survived and became famous is that most functions work via *networks* that spread through different locations in the brain. Most of these have some redundancy, but many networks also have specific critical sites that are necessary for normal working of that particular task, such as movement, vision, or memory. In addition, the brain demonstrates plasticity; when a function is lost, other areas of the brain can, to variable degrees, change to take over the lost function. Therefore, Phineas did lose some function—he became less motivated and his behavioral choices less prudent—but he also demonstrated recovery, though to an incomplete degree. This is consistent with the location of his injury and what we have since learned about brain plasticity, as well as the role of the frontal lobes in judgment, evaluation of choices, and reward circuitry. As our story progresses, this kind of example will become important for understanding the degree to which the reward system is malleable.

Critical clues about reward biology and decision making came from the study of Parkinson's disease. This common neurological disorder affects about 1 million people in the United States, causing impaired control of movement and disabling tremors. Parkinson's disease results from gradual degeneration of the neurons in a tiny area in the brainstem called the substantia nigra. Advances in light microscopy at the end of the nineteenth century enabled researchers to identify that this site was damaged, but they didn't know why damage here might cause the symptoms that patients suffered. It was the seminal discovery in the 1960s that the brain communicated mostly via chemical neurotransmitters, and that decrease in the transmitter dopamine in the substantia nigra was responsible for Parkinson's disease, that provided a major leap in understanding brain function and garnered a Nobel prize.[5] This very focal deficiency of dopamine in the substantia nigra resulted in severe disability or death until researchers discovered ways to increase brain dopamine with medications that could restore some of the lost function, sometimes dramatically "unlocking" patients who had lost the ability to move or talk.[6] But it was the unexpected side effects of some of these miraculous dopamine-restoring treatments that led to serendipitous insights about reward circuitry.

The surprising observation was that treatments to increase brain dopamine activity with medications or, later, electrical stimulation, caused strange and troubling behavior in some patients. Even people who had never had these tendencies before might develop a new obsession with gambling. Some went on serial shopping sprees, developed hypersexual behavior, began collecting obsessively, or showed other manifestations of severely impulsive, goal-directed, and out-of-control behavior. Even compulsive singing was reported.[7] Some developed a kind of addiction to the dopamine-increasing medications themselves. What did any of this have to do with the motor circuits, or with the functions of dopamine?

To solve this mystery, the neural basis of psychological functions had to be better understood. The key decision-making functions employed in gambling—indeed, in most decisions, as we will see—include assessment of risk weighed against the experience of reward. Pathologic, compulsive gambling means that a person's ability to weigh these factors is somehow unbalanced. Since Parkinson's treatments work by increasing brain dopamine, this was thought likely to somehow underlie the strange behavior. Could it be that increasing dopamine in the damaged substantia nigra restored an imbalance caused by the disease, but "overdosed" other undamaged dopamine pathways, hyping up how rewarding it feels to win at gambling? Or did patients somehow fail to perceive or care about the risks of losing their money?

The mysterious link between Parkinson's disease, dopamine, and impulsive behavior drew in researchers from many fields within brain science working in parallel. By pooling data on thousands of North American Parkinson's patients, neurologists discovered that over 13 percent of patients had compulsive gambling, buying, binge eating, or hypersexual behavior. This subset of patients had some features in common: they were treated with dopamine agents, tended to be younger, male, live in the United States rather than Canada, often were smokers, and had a family history of gambling. This suggested that the dopamine treatment triggered something in patients who might also carry some genetic and environmental predispositions.[8] Behavior is rarely simple—it is almost always a dance between the brain you have right now, molded by your experience and your inheritance, interacting with the world you encounter at this moment.

Invertebrates, rodents, and humans have major differences in how their brain systems are organized for learning, but there are enough similarities that as research subjects, they all can contribute to revealing basic principles. Using a different approach to analyze the Parkinson's patients' strange behaviors, research psychologists taught rats to "gamble" and then tested the effects of deep brain stimulation in the main targets used for electrical stimulation to treat tremor in human Parkinson's patients.[9] How in the world do rats "gamble"? These animals learn complex choice experiments involving food reward with amazing alacrity, reflecting their sophisticated reward circuitry. A gambling paradigm allows the rat to choose between pushing different levers to get food pellets. Each lever has a different chance of shutting off for a period of time after the rat pushes it. Using trial and error, the rat figures out which lever tends to behave which way, and then decides how much of a chance to take to optimize the amount of food it earns—a veritable rodent slot parlor. If the rat pushes a lever too soon, it is penalized with a longer delay until the next possible reward. By stimulating the part of the brain targeted in Parkinson's patients, researchers observed an increase in the rats' tendency to respond impulsively (push the lever too soon) when there were multiple choices available, but not when the rat just had one choice at a time and a big "time out" penalty for a mistake. These findings were consistent with some of the features seen in Parkinson's patients who behave impulsively and take bigger risks when punishment for an ill-advised choice is relatively mild.[10] Thus the reward system isn't "all or none" like a toggle switch—it is designed to be modulated to work in particular ways under particular circumstances.

In other centers studying the problem, patients with deep brain stimulators who developed pathologic gambling helped parse out the role of specific genes and different sites within the reward network.[11] Elsewhere, during surgery to place deep brain stimulators in patients with depression or obsessive-compulsive disorder, surgeons recorded the activity of individual brain cells from a critical reward circuit site called the nucleus accumbens while the patients, who were awake during surgery, performed a computerized card gambling game. Neurons here turned out to be specialized to assess the difference between the rewards subjects expected and what they actually got, an assessment that may be altered in patients with pathologic gambling.[12] On other fronts, evolutionary behavior scientists and market experts studied the role of hormones and risk-taking in healthy college students and in pathologic gamblers, illuminating other factors that might explain why some people—especially younger men—might be more likely than others to experience this side effect.[13] Other researchers discovered increased risk-taking behavior in patients with damage to other brain regions, including the frontal lobe, the site that was damaged in Phineas Gage.[14]

With these diverse approaches, these strange behaviors involving disruptions in the human reward system now have been worked out to a significant degree. Many factors play a role—some common biologic tendencies, influenced by genes, hormones, circumstances, and history—and this behavior can be modulated. All of this is good news, because pathologic gambling and other impulsive behaviors in Parkinson's disease are now better understood, more easily prevented, and more effectively treated. The pathologic gambling story shows how something unexpected in a disease process can shed light on how the brain works to make decisions under normal circumstances, and how behavior involving decision making and consumption is neuronally organized and modified.

Imaging Tools: New Views in Reward Biology Research

Magnetic resonance imaging (MRI) uses magnetic fields to temporarily align molecules in tissues and then detects the magnetic waves they give off as they "un-align" over time. Tiny differences among tissues generate a detailed three-dimensional image. A typical brain MRI involves multiple different magnetic pulse sequences that each influence the tissue in different ways; the vibration of the magnetic coils is what makes the loud noises you hear when you undergo a scan, and why it often takes a long time. Radiologists have learned the patterns that mean that a tissue is one thing and not another, and can tell the difference between, say, a blood clot, a tumor, and an infection.

The MRI revolution has provided tremendous insights into the structure of the brain as well as how its parts are connected and how it works. One type of MRI scanning used extensively in reward studies is *functional MRI,* or fMRI. When a particular part of your brain is active, blood flow to that area increases. These tiny differences in blood flow to particular regions can be turned into a picture, which shows the active area "lighting up" compared to less active areas. In this way, parts of the brain that are involved in certain motor functions or cognitive processes can be visualized. When surgeons plan to remove a tumor near the area of the brain that controls a patient's hand, an fMRI obtained while the patient taps the fingers together can show where that critical area is located in relation to the tumor, helping the surgeon to avoid it during surgery. While functions tend to be located in certain areas, there is enough variability among individuals that knowing the exact location for the patient is extremely useful.

The same principle can be applied to thought processes like language. In the scanner, you might be asked to think of all the words you can recall that start with the letter "A." The relevant parts of the language network can be identified by tiny increases in blood flow during the mental task, and this can be turned into a picture with those areas displayed, for example, as yellow on a gray background of brain.

In research investigating the human reward system, subjects can undergo fMRI studies while performing mental activities, such as decision-making tasks, placing bets to obtain cash rewards, or playing video games.[15] A related method is resting-state MRI (rsMRI), which measures correlations of brain activity in different areas to see what functions might be linked, and therefore what parts of the brain might be functionally connected. PET (positron emission tomography) and SPECT (single photon emission computerized tomography) utilize nuclear isotopes that are concentrated in an area that is metabolically active.[16] Many different manipulations can be done to tease apart the complex interactions of different brain structures, networks, and external circumstances that might influence reward. Imaging has been used to study the effects on the brain of music, meditation, drugs, social influence, empathy, decision making, placebos, fear, stress, parent-infant interaction, and many other cognitive and emotional processes, including those related to environmental issues. Much of the information we will review about reward biology in upcoming chapters comes from this type of research.[17]

All of these studies, though, have errors.[18] Sites in the brain don't actually "light up"; rather, these images are graphics to display very small, computer-processed changes, and there remains some controversy about the reliability of interpreting imaging findings regarding behavior and psychological functions. Nonetheless, many scientists depend on these techniques to provide information about reward system biology in living humans.

Brain Cells at Work: Added Insights from Single-Cell Recordings

We now have seen how imaging shows brain function at a regional level. Next, to understand how reward, learning, and decision making work at a cellular level, scientists have used microelectrode recordings from single cells to learn what specific job individual neurons are designed to do. These advances have been important for understanding how the system is designed to function and all the factors that play into neural events like making decisions. For instance, in vision, some cells respond to lines and orientation, some to color, and some to movement. These cells converge their knowledge on successively higher-order processing cells to allow them to recognize complex patterns like human faces. Looping in memory allows nearly instantaneous recognition of whether the face belongs to someone familiar—a useful skill for survival.

Similarly, recordings from thousands of cells reveal their individual roles during learning and decision-making tasks.[19] One of the main mechanisms for how the brain uses reward to help you learn arises from specific dopamine neurons whose task is to signal a "reward prediction error." The term "error" here doesn't mean you made a mistake. Rather, these cells increase their activation rate, and thus release more dopamine, when *unexpected* good events occur. This teaches the organism to learn a beneficial association, like a sound cue preceding a desirable food reward.[20] Reward-activated dopamine neurons release a phasic burst of dopamine lasting a fraction of a second, enabling critical neural tasks that allow you to perceive, decide, act, and adapt to changing circumstances.[21] This happens when rewards are greater than predicted during simple *stimulus-response conditioning* (a bell means food is coming), in *instrumental learning* (if an animal performs an action, like pressing a lever, a food pellet drops into the well), and in *reinforcement learning* (the rat learns to navigate a maze through successive associations of a particular "state" with an expected value).[22] In the intricate and super-specialized ways the brain is designed to perceive and process information, other dopamine neurons respond specifically to the relative importance, or salience, of a stimulus, such as how likely it is to signal a reward or punishment. Still others respond to the degree to which something is aversive.[23] We will delve into how different behavioral choices are evaluated and decisions are made, mechanisms whose elucidation also relied heavily on single-cell recordings, a bit later.

Complementing the single-cell recording approach, second-to-second changes in neurotransmitter levels in particular locations can be tracked by electrochemical probes while subjects receive food rewards, solve problems, encounter other animals, receive addictive drugs, and undergo deep brain stimulation.[24] As one attention-getting example, researchers studied what happens to dopamine in a part of the reward network when a male rat encounters—and mates with—a receptive female rat.[25] As in many rewarding behaviors, it turns out that the anticipation releases as much or more dopamine than the thing itself. (They didn't measure what happens in the female's brain during the same scenario; one can only speculate.) Dopamine-based rewards are designed by evolution to be brief and transient, a relevant fact for understanding many behaviors with environmental consequences, as we will see.

In humans, direct recording from inside the brain is restricted to patients who have electrodes implanted to treat a medical condition, but with patient consent, researchers can take advantage of these situations. In patients with Parkinson's disease, real-time dopamine measurements were made in the striatum, a location important in reward circuitry, while patients played a computerized money-investing game involving betting.[26] These patients acted similarly to animal subjects in the "rat gambling" tasks and to humans tested in other ways—but with a subtle difference. When the patients made financial investment choices, their dopamine release in the caudate nucleus did not reflect simply whether the reward they got was greater or less than predicted, as has been typically found in animal "gambling" studies. Instead, the dopamine levels reflected not just what the subjects got for a reward, but also how this differed from the best or worst the reward might have been, had the subjects made another choice. While this could reflect something specific about patients with Parkinson's disease, it may be that only humans can reason through the options to feel worse when they get something—even something better than predicted—but might have gotten even more. It's tempting to wonder if this could make us more greedy and envious than other animals!

How have all these levels of understanding—from human diseases, imaging, and single-cell data—come together to create a working diagram of the human reward system? Addiction is a major health problem involving perturbations in decision making, reward, and consumption. Thus, some investigators have postulated that clues from addiction are particularly relevant to our climate crisis. Research in this area, which integrates all of these approaches, provides our final medical example before we outline how the normal system is thought to work and the factors that weigh into our decisions.

Addiction: Reward Biology Gone Awry

If reward is a normal, "healthy" function of the brain, addiction occurs when one kind of reward, elicited by a drug or other kind of stimulation, becomes the predominant driving force for an individual's behavior, even if it is self-destructive. Some commentators have likened our climate crisis to a planet-wide, continuous low-level "addiction" to consumption and the status quo of a fossil-fuel way of life, even though it is destructive in the long run, and have looked for signatures of addiction-like overconsumption in our "normal" behaviors.[27] To assess this idea, we can look at the evidence for behaviors that may more directly fit the definitions of addiction in the clinical setting and their possible neural underpinnings. So-called "behavioral addictions" in humans—those not involving drugs, but instead reflecting compulsive activities such as gambling or shopping—have helped bridge the gap between addiction and human consumption behavior.[28] How and why do people get addicted to a behavior, and do these mechanisms have relevance for "normal," or what some would call the "normalized excessive," consumption of much of the high-income world?

Addiction disorders have major impacts on health and society, and their biological correlates have been the subject of a large body of research. While there are many theories about both the biological and social reasons for addiction, most researchers postulate an interaction between these influences.[29] Our reward systems are so universal and evolutionarily old that crayfish, fruit flies, worms, and honeybees can be addicted to the same substances as humans can, and serve as neural models.[30] Research showing how substances change the genetic instructions in specific brain sites has been cited to help explain why addiction is so hard to overcome.[31]

As we've seen, the reward system is not designed to make you "feel good" overall—it's designed to help you learn what you need to survive and thrive. Addictive substances generally work by accentuating what feels like "normal" reward—at first. They do this by pharmacologically making the normal reward mechanism go into overdrive; opioids and many other substances of abuse work by enhancing the actions of dopamine. But as we've seen, with normal learning, over time the reward prediction error signaled by dopamine decreases once the association has been learned, freeing you up to learn new things. But with addictive drugs, the dopamine surge remains artificially boosted long after the learning of reward associations has occurred.[32] The brain tries to get things back in balance by compensating for this increase in reward-related chemical release—it has many beautifully designed adaptations to try to accomplish this, to keep you healthy and functional despite changes in your internal environment. But over time, some of these changes backfire. The drug itself gets less rewarding, but going without the drug feels worse and worse.

There are multiple theories of addiction among neuroscientists, and many approaches to addiction link the changes in behavior to alteration in functions of specific brain networks. In one theory, those parts of the circuit that are activated by the learned associations with the drug—people, places, things—increasingly direct behavior not by conscious control, but by more automatic *habit* behaviors.[33] Habits are neural network patterns that are designed to be firmly strengthened and linked by repetitive association with reward.[34] Other researchers make a distinction between habits and goal-directed behavior, as addiction has elements of both.[35] Evolution designed habits resulting from repetition ("overlearning") to be deliberately difficult to break, because they represent important lessons taught by the individual's experience that usually provide a survival advantage. But in addiction, these kinds of padlock-strong associations, stamped in by the unnatural reward value of the drug, are counterproductive. Smokers who have tried to quit can articulate this quite vividly. Many associations of things, places, and people can trigger the urge to smoke, long after the conscious mind knows it would be a good idea logically to alter behavior—and in fact, long after the drug itself has been completely eliminated from the body. At that point, it isn't the drug itself that is causing the tendency to relapse—it's the linkages in the brain that the drug created in the past but are still present. This is why many addiction programs use the term "in recovery" for patients who were addicted in the past; it's a nod to the fact that some addiction related reward system changes may be long lasting or even permanent. Cravings represent unnaturally strong associations in the brain that are very difficult to "un-wire"—this is in part what makes some substances particularly addictive.

To be classified as addiction, the substance or behavior has to be initially rewarding, more of the substance or behavior is needed to get the same effect over time ("tolerance"), and the person spends increasing amounts of time in addiction-related behavior to the exclusion of other activities—even when this may be harmful to life goals or relationships. There is a sense of lack of control or a compulsion to seek the addictive substance or activity, and there are negative consequences to stopping the substance or behavior ("withdrawal").[36]

Can behaviors like gambling or shopping be "addictions"? Here, scientists have used behavioral criteria to define whether a behavior, rather than a drug, can be truly addictive—and have amassed evidence for phenomena like video game addiction as meeting these criteria. The first reports of patients exhibiting video game obsession with features of addiction began to appear in the 1980s, and scales to test the degree of addiction in gambling and newer types of compulsive behavior were developed for use around the world.[37] Like other behavioral addictions, habitual video game playing, which most often affects adolescent and young adult males, can become compulsive, all-consuming, and deleterious to normal life activities. Even things as basic as eating, sleeping, and caring for children can be overridden by gaming addiction. Individuals who are addicted suffer intense anxiety and craving when they are kept from engaging in gaming. The disorder has been reported to occur more often in children with attention deficit hyperactivity disorder (ADHD), people with histories of other kinds of addictions or social problems, as well as patients with Parkinson's disease treated with dopaminergic agents and who, as we have seen, are predisposed to gambling addictions.[38] We will explore why video entertainment in particular may be so rewarding from a biological perspective in Chapter 5.

As the phenomenon increased in frequency globally, neuroscientists began to study video game and internet addiction with brain imaging tools. Dopamine release and other reward circuitry changes were measured during video game playing using PET scans, changes in the volume of specific brain regions in gamers compared to controls were discovered, and changes in gamers' brain activation patterns were seen using fMRI.[39] While these types of studies carry many caveats, most studies have shown alteration in reward system function that suggests a biological signature of problematic behavioral addictions. Taken together, studies of addiction—the reward system when it's "hijacked"—have been one of the main sources of information about how the system is organized and what influences its function in real life situations.

We now have an understanding of some of the clinical contexts, tools, and intersecting approaches used in brain reward research. With this background, we are equipped to explore how the human reward system is constructed and how it assists learning and decision making, in particular with respect to choices about consumption and climate change. We will then be ready to look more at how the system operates in present-day humans, how it interacts with remarkable alterations that have occurred in how we live in recent human history, and how the brain itself can—or can't easily—be changed.

How the Human Reward System Works

Brain researchers describe the "reward system" in different ways, including a varying list of structures and different definitions of "reward." One typical description of a reward is something that "promote(s) a motivational state to allow an organism to purposely select the best way to achieve the successful procurement of its essential needs."[40] This kind of definition is most applicable to those studying the reward systems of animals, in which effort (motivation) to exhibit goal-directed behavior in order to obtain a reward (such as food or an addictive drug) is manipulated and studied. In humans, the concept of reward can take on additional nuances, such as whether specific brain circuits are activated by something that is important for functional or survival significance ("salience") compared to something that is perceived as "pleasant."[41] As our story progresses, we will see how the design of this system affects our behavior at a large scale to contribute to our environmental quagmire.

So, what do we mean by a "system"? In physiology, this term is used to refer to a collection of anatomic elements that work together to serve a particular function or purpose, akin to the heating, electrical, or plumbing systems in your house. Animals have systems for breathing, circulation, digestion, movement, and nervous control. Likewise, within the nervous system are separate systems for different brain-related functions, such as the visual, auditory, motor and neuroendocrine (hormone) systems, all of which are made of networks of neurons and their targets. Systems often include feedback loops, widely dispersed inputs, cross-talk between different networks, and widespread outputs, so that the systems are coordinated among themselves, like the way the thermostat in your house may have a role in both the heating and plumbing systems. Thus, an individual neuron or pathway may be used for several different systems as part of the overall design.

Concepts of the design of the human reward system have evolved with new discoveries like those described above. Many people have heard of the "limbic system"; this term is in wide use, although the original theories about the interactions and functions of its components have been amended with more recent insights. Originally, these structures, which curve around as a sort of "limbus" or border to some of the central structures of the brain, were thought to be the main system subserving emotion, along with some additional functions, such as olfaction. The link to the sense of smell is often touted as the reason that smell evokes such powerful and often emotional memories.[42]

It is now recognized that the classic limbic components have more widespread and complex interactions than just controlling "emotion," including major interconnections with the reward system, and that together the limbic and reward-system structures work in concert to support learning, affect, motivation, decision making, and related cognitive processes.[43] The paired curlicues of the limbic structures sweep around the brain's central fluid-filled ventricles from the front of the temporal lobes, about an inch deep to your temple on each side. They include the *amygdala,* an almond-shaped nucleus related to fear processing; the *hippocampus* ("seahorse"), which is critical in memory functions; the long, curved *fornix* ("arch"), which is the main outflow highway for the hippocampus, the paired *mammillary bodies* in the hypothalamus, and several other interconnected locations (Figure 2). To early brain scientists, the concept of a circuit was particularly intuitive for these structures, because their linked shapes make an easily seen open circle that can be recognized with the unaided eye. This striking appearance may have contributed to earlier ideas about their potential function, when other interrelated parts of the brain that were less visually obvious remained obscure. At present, many authors describe the interplay of parallel circuits known as the mesolimbic system with those of the "historical" limbic system as making up the integrated "reward system."

One such structure that turned out to be critical to the whole way reward works was hiding in plain sight for generations of brain scientists, with functions that were obscure. It doesn't have the beautiful sweeping shape of the limbic system, and in fact it's hard to really see as a separate structure at all. Even most neurosurgeons can't really describe where it is. But for all its ordinariness, it is exquisitely designed to work from a location that is ideally suited to the principles of neural design to optimize efficiency, efficacy, and flexibility.[44] This structure is the nucleus accumbens (Figure 2).

Even its name sounds insignificant—it means "nucleus next to." Its full name, nucleus accumbens septi—"nucleus next to the septum," a membrane that marks the midpoint between the two halves of the brain—is almost never used. So, the nucleus accumbens was left dangling, in name and understanding, until advances in science helped uncover its amazing and central role in reward circuitry.

The next key reward system components are two tiny nuclei in the midbrain segment of the brainstem: the *ventral tegmental area* (VTA) and the *substantia nigra pars compacta* (SNc). We met the substantia nigra in our discussion of Parkinson's disease. The names reflect their "address" in the brain based on standard anatomic planes (Figure 3) and / or what they look like under the microscope; most brain structures were named based on their visible features long before anybody knew what they did. These little spots are the main sources in the brain of the crucial neurotransmitter we met previously, dopamine, that is used in both reward and movement. Medical students tend to remember these structures because in midbrain cross sections, they look strikingly like a face; the SNc, which is the part of the brain affected in Parkinson's disease, is shaped like paired eyebrows on either side of the v-shaped, "bridge of the nose" VTA, and above the easily seen "eyes" that are the red nuclei (involved in motor coordination) (Figure 4).[45]

The relationship between these little spots and the rest of the brain illustrates the design genius wrought by evolution to enable you to learn, adapt, and function in changing circumstances. In general, the VTA interacts more with reward circuitry, particularly in the nucleus accumbens and related sites, while the SNc interacts more with the motor system via the basal ganglia. There are advantages to this distribution of labor, as we shall see. The connections between the VTA and the nucleus accumbens help make up the "mesolimbic" circuit, which is essential for reward function; "meso" in this context refers to the midbrain portion of the brainstem, where the VTA and the SNc live. Besides this specific mesolimbic loop, both the VTA and the SNc have widespread interaction with and feedback from a diversity of brain networks and with each other. There are extensive interactions with broad areas of the cortex, especially to provide input into the decision-making areas of the frontal lobe and other networks; for this reason, to be thorough, some workers use the mouthful-of-a-term mesocorticolimbic system to designate the entire reward system.[46] Though small, these dopaminergic spots in the human brain have more cells than do those in other animals—more than half a million—and are extremely well connected to sites throughout the brain, incorporating all kinds of intersections and feedback loops.[47] This helps explain how important the reward system is to teach you what you need to know to optimally navigate

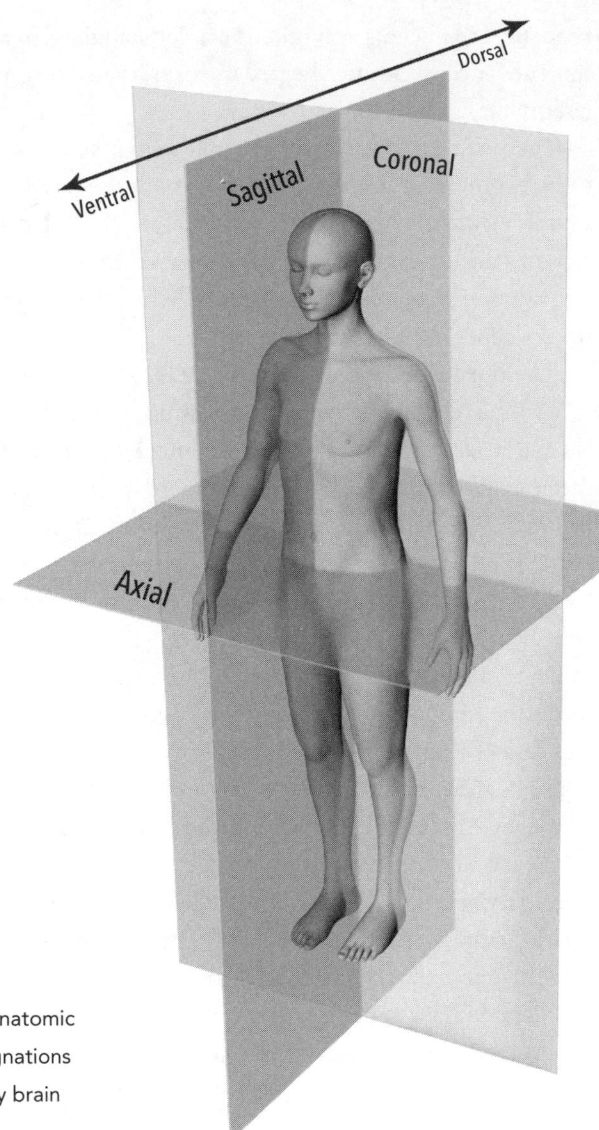

Fig. 3 Diagram of the standard anatomic planes and direction designations used in the names of many brain structures.

in the complex, varied, and ever-changing world inhabited by each individual and unique human being.

That reward and movement might be linked makes perfect sense, because when you need to learn that a particular behavior results in a reward that is beneficial for survival, the reward system has to teach the lesson and the motor system has to enact—and repeat—the behavior to be learned that is associated

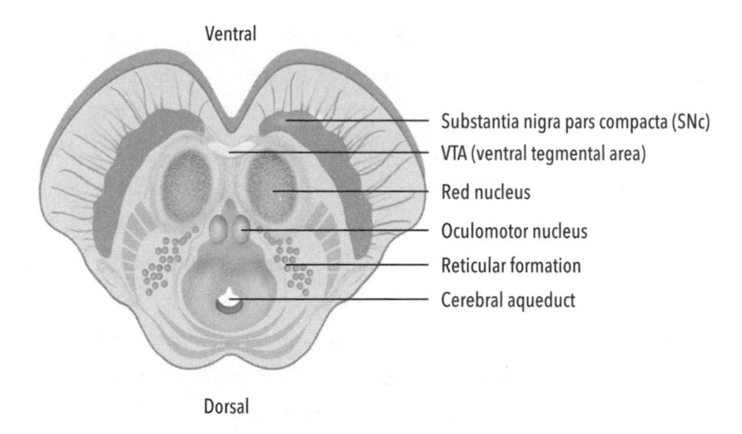

Fig. 4 Cross-sectional diagram of the midbrain. The main sources for the neurotransmitter dopamine, central to the function of the reward system, are located in the two small nuclei shown here, the ventral tegmental area (VTA) and substantia nigra parts compacta (SNc). This brain region is often described as looking like a face; the "eyes" are made by the red nuclei (important for movement), the "eyebrows" are the SNc, and the "bridge of the nose" is the VTA. The "mouth" is the cerebral acqueduct, through which cerebrospinal fluid passes.

with the reward. Picture a baby figuring out, with increasing efficiency, how to pick up and eat finger foods. Thus, these two nuclei, which are close together, both act via releasing dopamine (and some other transmitters, like glutamate, that are helpful for learning). They are exquisitely responsive to second-to-second changes in the internal and external environment, and coordinate their interactions with other parts of the system. The fact that the neurons from these tiny spots can make millions of instantaneous connections is why, after you slip on an area rug, the next time you walk down that hallway you'll tread more carefully, without even being aware of it.

When dopamine is released from the axons of specific neurons, it works by "smoothing the connections" between different loops in the system to make them more likely to fire together in response to a particular stimulus as a result of learning (e.g., "restaurant" = "food"). It also is the means by which the brain tags something as being rewarding ("French fries" = "good").[48] Single-cell recordings have elucidated the clever computing by which this is done. Specific dopamine neurons involved in reward processing fire in proportion to whether the utility of something you encounter—the subjectively determined usefulness of that thing at that particular moment—is greater than predicted. The *reward prediction error* is independent of the specifics of the sensory characteristics of the trigger; instead, it reflects how much the stimulus is valued to the animal or person at that moment. To accomplish this complex task, different neurons throughout the brain are specialized to respond to specific features of their input, which includes thousands of synapses for every cell. All of these connections enable the dopamine neuron to receive outside sensory input and to sense the organism's internal state, as well as to access information from memory. These complicated network computations allow the animal to recognize a potential reward, weigh the value of a particular choice at a given moment, translate this into action, get feedback on the consequences, and teach and store the lesson for future use.[49] Neurons often follow the same computational formulas used for assessing value in the field of economics, and these same mathematical principles have been applied successfully to machine learning.[50] These parallels will be relevant as we compare the brain-based "utility calculations" when available choices might include potentially pro-environmental behaviors.

Mesolimbic reward structures interact in ways that make sense for how things need to function. In particular, the paired hippocampi in the front of each temporal lobe are critical for stamping in new memories (see Figure 2). In most right-handed adults, the left hippocampus processes most language-based memory, while the right subserves visual and spatial memory. In the reward system, these structures and their connections help you to remember what the other parts of the reward system designated as "rewarding," so you can act accordingly next time you meet that reminder. Restaurant! French fries! Yum!

The fingertip-sized amygdala sitting right in front of the hippocampus is important for learning memories associated with emotional events and for processing fear, aggression, motivation, and reward.[51] It plays a role in enabling empathy and trust, as well as determining where a risk is located in space.[52] Like the surface sensors on bacteria we met in Chapter 1, the amygdala helps arbitrate when both a reward and risk occur simultaneously, such as foraging for food under dangerous circumstances, and it facilitates learning the association when something negative occurs.[53] This part of the circuit likely was involved when your brain intervened in tiny, behavior-tweaking, below-your-conscious-awareness ways to deflect you from striding too carelessly on the loose area rug once it had learned that lesson.

Making Decisions

The front part of the brain's neocortex, or outermost layer, is a crucial part of the reward system, involved with making assessments, judgments, and decisions. It does this by learning what sensory inputs are associated with "rewarded" results, what behaviors lead to what consequences, and then factoring this into making a decision about how to react when faced with a choice. This is not all—or even mostly—at a conscious level. While most people can give some explanation for why they made a particular decision, the sheer magnitude of tiny influences on each behavioral choice makes even an introspective accounting incomplete. What goes into determining how we learn which choices to make, and how might this influence those decisions that affect environment-related behavior?

Let's begin in the frontal lobe, which extends from behind your eyebrows to the top of your head. Phineas Gage's tamping iron rocketed through this area. The "prefrontal" region denotes that part of the frontal lobe nearest the forehead. Different parts of the prefrontal cortex integrate different aspects of decision making; the medial (closer to the middle) prefrontal cortex is involved in motivational behavior—what it is that makes you get up and do something, while the lateral is more involved in behavioral control—what keeps you from doing something impulsive or acting too soon.[54] Damage here explains Phineas's altered behavior.

The *anterior cingulate cortex,* arching from front to back just over the white matter connections between the two halves of the brain, links the value of a reward to decisions about the degree of "cognitive control"—the assessment of self-control appropriate for the circumstance (see Figure 2).[55] How much risk you'll take to get food depends, for instance, on how hungry you are, whether you're responsible for feeding your offspring, what the risk is and your prior experience with it, how much and what kind of food can be gained by the effort, and numerous other factors you weigh based on your experience and your motivation level to make a decision about a given action at a given moment in time.[56]

One of the important characteristics of the role of reward in decision making is that the reward value of a particular thing is not a fixed, intrinsic property of the thing itself. Instead, the reward value of something varies as a function of the state of the animal or person at the time the potential reward is encountered, as well as prior experience with that thing.[57] Shifts in whether something is rewarding or not can be dramatic. Think of how a tasty snack food makes you groan after you've overindulged. As a more extreme example, one notable brain imaging study showed that patients with anorexia nervosa experience visual representations of doing without food as rewarding.[58] This is a complete reversal of the normal reward value of something as basic to survival as food.

How does the brain decide among possible behavioral choices to whether one path will lead to more reward than another? It is endowed with a marvelous computational system whose function is to perform this critical evaluation. Specific orbitofrontal neurons in the prefrontal cortex fire in response to the relative "rewardingness," or utility, of different choices, varying their firing rate when a reward becomes less rewarding during satiety or other circumstances that lower its value.[59] This means that the cell has to get input from thousands of other neurons in order to make such a sophisticated determination. The same cell will fire in response to a different reward altogether—its job is to process and signal to the other parts of the system that ultimately decide whether to work for more of a given kind of reward (in primates, for instance, drops of juice or a small piece of banana). The cell doesn't say "this is juice" or "this is banana"—that's the job of other cells. Rather, it says "this is how useful or good this thing is to you at this particular moment, based on all the inputs now and in your past experience." This hierarchy of preference is manifested in the animal's choice during a behavioral task. Once the animal chooses and obtains the reward, the cell responds to that event as well, and can change its response over time as circumstances change, such as the availability of new kinds of rewards, or a penalty associated with the reward. This kind of neural arrangement is extremely efficient, because the animal's brain can converge all kinds of internal and external information to make utility judgments about both familiar and brand new rewards in their infinite variety. It can update its valuation of a given choice continuously, allowing for remarkable flexibility and adaptability. Other cells in other locations code for which choice is least preferred, how much motivation the animal shows to obtain that reward, and innumerable other factors that might weigh into a decision.[60]

A cornerstone of so-called "economic choice" behavior and its neural underpinnings is that there is no "correct" choice. Decisions are made by subjective valuations, that change with accumulated history and present circumstances.[61] Our valuations of specific behavioral choices, while tending in certain directions much of the time, are designed to be changeable. In addition to immediate choices with only short delays until the goal is realized, the brain is equipped with some ability to think ahead to rewards we might get in the future.[62] This flexibility can take into account even novel threats, like climate change, but the strength of its relative weight in decision making depends on many factors, as we will see. Our politician deciding on how to vote on the climate bill will be influenced by all these minuscule influences, and we will learn more about factors specifically influencing political decisions in Chapters 3 and 8.

Some authors, building on concepts described by Sigmund Freud, have popularized the idea that decision making is performed by two "brain systems"—a more reflexive, automatic, unconscious system that makes snap judgments based largely on associations and heuristic shortcuts, and a slower, more conscious, rule based, rational, and deliberative system.[63] This concept is useful for differentiating two general contexts of cognitive events and "thinking" as we experience them subjectively—and to point out how people can get it "wrong" in their initial assumptions and decisions about choices they encounter. But in neural terms, brain function is much more complicated than two mutually exclusive and distinct systems. As we've seen, literally millions of events are happening simultaneously, in countless parallel and overlapping and interdependent networks, ricocheting all through the neural pathways at any given instant. Even when we think we are being deliberative, myriad obvious and subtle inputs and interactions are working to influence the outcome of our deliberations, in time frames that are not cleanly characterized as immediate or gradual, nor with only two systems involved—almost every part of the brain gets into the act. Nonetheless, common experience shows that we can "jump to conclusions" or "think things through"—and these distinctions have intuitive and pragmatic validity. But the brain does not have a switch; it is never turned off under normal circumstances, and neural processes are more a continuous but highly organized and functional maelstrom than a series of time-locked and separable occurrences.

The science that has unraveled how the human reward system works shows that it is not reductionist and deterministic; it is not entirely hard-wired nor totally predictable from genes. Rather, it is fluid and interactive with what we encounter in the world.[64] The scientific understanding of reward is not incompatible with other ways of conceptualizing "higher" human traits and values—altruism, courage, appreciation of beauty, self-determination, spirituality. But it holds to the premise that the many variations of "human nature" can be described in the language of things that happen "inside the brain"—with the full understanding that what happens inside the brain happens in concert with what the person interacts with in the external world. Each of these parts of how the system is designed plays into our minute-to-minute decisions, and those of a world full of people making choices at multiple scales that may influence the future of climate change.

Let's conclude by watching the brain in action on an ordinary day.

Lunch Break

Imagine a regular day at work. We'll assume you work daytime hours. You always eat at the same time every day, let's say, 1 P.M. This morning you were running late, so you skipped breakfast. At around noon, without conscious effort, you start thinking about food. The word "roll-out" in a document you're reading conjures for a brief instant an image of your favorite dinner roll, warm and buttered. You find your mind wandering to French fries without particularly knowing why, but this happens right after you handle a manila folder which is fry colored. You put off a task that will take 30 minutes to complete, in favor of one that you can wrap up in 10 minutes, which involves stopping by the desk of a colleague who's often up for a quick trip to the coffee shop. What's happening in your brain?

Think of your nervous system as billions of tiny, wispy, sparkly cells, with long, feathery arms and fingers reaching out and touching thousands of other cells. Each delicate, complicated cell in this microscopic universe has a job to do. As you are trying to concentrate on your work, cells in your hypothalamus, a specialized area in the base of your brain, are sensing your blood sugar level to see if it's low, and others are checking your salt level. Other cells reach long arms into your stomach to see if it's full, into pressure sensors in your neck to see if your blood pressure is dipping, and simultaneously thousands of other physiologic parameters are being monitored continuously. The most complicated bank of equipment in the most sophisticated hospital intensive care unit can't hold a candle to what your brain does automatically to keep you healthy as the world inside you and around you changes every second. These monitoring cells send the information to multiple sites, including your dopamine networks, which in turn signal your nucleus accumbens and the various segments of your prefrontal and anterior cingulate cortex, which each have a specialized role in weighing the pros and cons of each possible decision on a second-to-second basis.[65] When your body senses hunger through its network of hormones and receptors, the feedback makes those things you have learned to associate with food more strongly rewarding.[66] At the same time, these areas also receive input from your hippocampus, which helps access the fact that at a staff meeting yesterday, your boss stressed the need for your division to speed up the reports for your current project. Cells in your hippocampus and amygdala provide the keys to a lifetime of associations (do we expect to eat at the same time every day? Is lunchtime generally a good thing, or a source of ambivalence?), and little sparkles of electrical activity recall subconsciously that an interesting new coworker may be at your lunch spot today.[67] All the while, cells in your midbrain's dopamine centers—VTA and SNc—are spritzing their microscopic doses of transmitter in order to choreograph your behavior by assessing and manipulating anticipated and current rewards. They base their dose and distribution of dopamine "lubricant" on the instructions you inherited in your genes that were selected for survival advantage, modified, as they are designed to be, by what you have learned in your life.[68] Right now, they are actively rewarding thoughts of food, because they are being pushed to do so by their inputs telling them that your body needs nutrition.[69] Thus they monitor and in turn contribute to the traffic on all these infinitesimal electrical highways among the far-flung neurons sensing your internal and external world that provide continuous status reports to the dopamine control center. As your sensors send signals that your tank could use refilling, your dopaminergic cells start leaning on pathways that have been associated with food reward in the past; the thought of a warm dinner roll or crispy fries provides both a reward and a motivation to act, reflecting hunger-enhanced activity in different twinkling parts of the spider web network. But you are human and evolved also to be motivated heavily by social rewards—what your boss thinks of you, what your coworker might say, what fun you might have, or what troubles might come your way if your work isn't completed. Without your full awareness, your prefrontal cortex, bathed in thousands of signals from of all these effervescent pathways, is weighing its multitude of inputs along with what it can figure out about the potential tastiness of your lunch and whether it might be something new or something totally predictable.[70] It is your little bland-looking nucleus accumbens, weighing also the buzzing local connections, including the wider ventral striatum surrounding it, that acts as the translator that turns all of these diverse inputs into the motor output needed to take action.[71] And so, based on your recent and your distant history and your unique genetic predispositions, sooner or later you will pause your computer screen and head out your door. How long you stay away, and what you do tomorrow, depends on the consequences of today's decisions along with the enormous databank from all your past days—those you remember consciously, and those you couldn't recall if asked but affected who you are today. Was the food particularly good? Did the coworker act friendly? Did you get your work finished on time despite an overly long but pleasant lunch break? These things will lead to tiny increases in the degree to which dopamine strengthens the association in the orbitofrontal cortex and nucleus accumbens between future memories of those circumstances and motivation leading to action. How all this affects you depends on the evolutionary influences on how your reward system is built and is modulated by your personal history, your genes, your gender, and many other factors that make you the unique person you are.[72] Did you spill your lunch on your shirt? Did your coworker make a snarky comment? Was your work sloppy because you took too much time at lunch? The amygdala and anterior cingulate cortex will get involved and sprinkle their transmitter or upregulate their receptor subtypes accordingly, in order to help predict the possibility of negative consequences next time you encounter a similar situation—and so to make the idea of "lunch break" a bit less rewarding than it otherwise might have been.

You are designed to learn. When the circumstances or the rules change, you are ready, and each new opportunity to experience something that defied expectations leaves a new mark to weigh in the next time you have to make a choice.[73] Events both positive and negative will make their mark via tiny changes in your brain cells, which will influence how your dopaminergic network tweaks your decision making tomorrow and in days to come.[74]

3

The Universe of Human Rewards

SOME REWARDS ARE JUST . . . HUMAN. Rats and cockroaches sport reward system underpinnings similar to ours, but humans stand alone on the podium for altering our planetary chemistry at a scale we now recognize threatens all living things. We are different in our capacities from other animals, and in our reward system functions. Because of humans' outsized frontal lobe, highly developed language skills, complex social structures, and capacity for adapting successfully to an enormous variety of circumstances and cultures, some rewards are unique to the human situation, so require other approaches to study them.

Along our timeline from Chapter 1, we're studying a biologic system that is built with equipment from our evolutionary past that spans millions of years but is being used for circumstances in modern society only a few thousand years old—the last few steps on our walk. Let's look more closely at the range of human rewards we experience and what's been learned about how the brain processes them. Because of their particular relevance to environment-related decisions, we will explore specifically the rewards of *money, prosocial interactions, agency, novelty,* and *familiarity,* as well as how cultural learning affects what we find rewarding. If climate change has occurred in part because particular human rewards have influenced our choices and behaviors, we will need to know to what extent things we find rewarding are "hard-wired" or can change, and if so, by what means.

Your Brain and Money

We'll continue our exploration with a "secondary" reward—something that's associated with the ability to obtain the "primary" rewards of food and other necessities of survival, and so becomes rewarding in itself. Money is the perfect example of a secondary reward. Its reward value is always learned; by itself, it isn't inherently especially appealing. But our brains are set up to learn to attach meaning to secondary rewards, just as the homonym "roll" and the color beige conjured thoughts of food when you were hungry during the lunch break in Chapter 2.

Money has provided a convenient subject for studies of reward in humans. It is readily quantifiable, has meaning to most people from childhood on, and conditions for obtaining and spending it can be easily manipulated in experimental situations. It often serves as a powerful motivator (for instance, in my practice, seeing if a child will reach for a five-dollar bill is one of my most reliable tests for emergence from coma). The desire for money beyond what is needed for immediate survival has motivated much of human history and led to discovery and progress but also has fueled behavioral choices and social and economic systems that have paved the path to climate change.

To study how humans interact with money, rather than looking inside the brain at the level of cells and networks, most research on economic behavior looks from the outside, at the observable decision making of human subjects in psychology experiments. Indeed, the behavior we observe exquisitely reflects all those billions of network connections and tiny neurotransmitter spritzes we now know underlie the valuation and choices made by each participant in a behavior study. By testing large numbers of subjects, tendencies of specific groups of people under specific circumstances to make similar behavioral choices can be uncovered. This can give us additional clues about how our neural equipment may be weighted to respond to different types of rewards in particular ways.

Many investigators have used games and game theory as research tools to study human behavior involving money as a reward. Classic economic decision games, including the Ultimatum Game, Dictator Game, and Prisoner's Dilemma and their variants, have served as the basis for much that has been learned about human behavior regarding money, negotiation, fairness, cooperation, and punishment in various experimental situations.[1] In the Ultimatum Game, Player A, the "proposer," is given a sum of money and decides whether to give some to Player B, the "responder." Player B can accept the offer or refuse it—in which case, by the rules of the game, neither player gets anything. In a purely rational world, the responder would always say "yes" to any offer, because it's more than getting nothing. But in fact, most often responders reject offers perceived as "stingy." This happens even when there's no opportunity for future reciprocity (that is, being "nice" to someone to increase the chance that the person will be nice to you when the tables are turned). For instance, if Player A starts with a hundred dollars and only offers Player B ten dollars, Player B may refuse, getting nothing, but ensuring that Player A loses all the money.

Researchers can vary all kinds of things in spin-offs of this kind of game—comparing players of different cultures or genders, looking at the effect of recent winnings or losses, manipulating the perception of one player toward the other player, and so on. Since much of this research is done with the subjects interacting with a computer screen rather than a real human, many characteristics of the interaction can be manipulated; often Player A is not an actual person, but is a "phony" persona, which ensures that the responder receives a set of predetermined offers designed to study the reaction to specific situations.

Another widely used tool for studying financial behavior, the Dictator Game, is similar to the Ultimatum Game but eliminates the second player's power to respond to the first player's offer. This variation focuses on understanding those factors that influence Player A to ever give any money away at all—since this isn't "rational" from a financial gain point of view. This can be used to compare how people from different backgrounds behave or the effect of manipulating players' immediate situation or perception of other players.

The third game, Prisoner's Dilemma, is used to study whether people tend to act in a purely self-interested way or to cooperate with another person if this leads to both people (or a group) benefiting even at a cost to the individual.[2] Such games have been used to predict the outcome of negotiations among countries or other parties where cooperation versus self-interest may be at play—including, incidentally, issues like climate change.[3]

Using these tools, economists and psychologists delve into many different factors that tend to lean people toward various types of decisions about money—both getting and giving. This body of game-based research has demonstrated that for most people, behavior about money and other similar resources is not fully "rational," in the sense that people do not always maximize financial gain over every other kind of reward. This insight led to major shifts in economic theory and helped to spin off entire subfields, including behavioral economics and neuroeconomics. It certainly has influenced marketing and advertising—figuring out how best to convince people to spend money on things that may not be in their financial best interest!

While these studies manipulate circumstances to bring out particular behaviors in isolation, as we have seen, in the real world people are influenced by many competing factors when they are making actual decision. Nonetheless, these kinds of studies have been carried out in isolated agrarian or hunter-gatherer societies, in young children, and even in chimpanzees, and no group yet tested has shown purely "rational" behavior with respect to money or similar resources.[4] This suggests there is likely something about our shared nervous system that tends to make us behave along similar general principles that we will outline later in this chapter.[5] Still, even though no group tested had a majority of individuals making financial decisions purely "rationally," there were significant differences among groups. Thus, there are common tendencies shared by most people, but there are also differences in behavior depending on life circumstances and individual characteristics. The brain is not deterministic—"forcing" people to behave a certain way. It has underlying mechanisms that create general tendencies, but these are interlaced with individual differences, both inherent and experienced. Because both our internal needs and external circumstances change from minute to minute, the decisions we make and what we find valuable also are changeable.

Beyond just observing how people behave, to find out what's happening in the players' brains during economic choice games, subjects can undergo brain imaging, electrical recordings, manipulations in brain chemicals by medications, or direct measurement of neurotransmitters in specific brain regions as they play, techniques we explored in Chapter 2. For instance, if you ask people why they reject stingy offers, most will tell you it's because the stingy offer made them angry.[6] Functional MRI (fMRI) performed during the Ultimatum Game showed that unfair offers elicit activity in limbic-connected structures in addition to the prefrontal cortex, which the investigators interpret as showing that unfairness elicits an emotional rather than purely rational response.[7] Because the stingy offer elicits anger, by turning down the offer rather than accept the measly amount, Player B, the responder, forgoes a chance to make at least some money, but also punishes Player A for being so ungenerous. It is in this way that purely monetary rewards can be influenced by a sense of fairness or by social rewards. These tendencies can be manipulated by altering the psychological state of the subject, such as whether the person is stressed or praised right before making a decision about how generous to be toward another person, or by giving subjects drugs that blunt some of the "anger" response to an unfair offer.[8]

But let's pause to ask a basic question about this kind of research. Does knowing what parts of the brain are involved when certain economic decisions are made offer insight into why we act the way we do? If we are not always rational about money decisions, and by extension, consumption in general, why is that, and what does it imply about the underpinnings of our behavior? Do we consciously deliberate with our logical, analytic, objective cerebral equipment, or are these kinds of decisions and behaviors influenced by our evolutionary past?

The science shows that, just as for more primary or basic rewards, those involving money also are characterized not just by spots that "light up" but by networks that deliberately strengthen certain neural patterns and weaken others. Behaviors don't just happen, passively—as we learned at the level of cells and networks, your brain is designed to respond to specific types of events, and to increase the chances that certain reactions happen in a weighted way. Recall that the designer was evolution, as expressed through your particular genetic brand, modified by your particular life history and present circumstances. Among those, the most weighty influence on how things tend to work in most people most of the time is the evolutionary history we all have in common. While the way we evolved to work in this regard generally favors survival of the individual, there may be some factors for the society overall.[9] This evolutionary undercoat means that it takes a lot of training, cognitive energy, and stamping in of alternate rewards to override the way the system is shaped by millions of years for survival—even for secondary rewards like money. Let's next look at how scientists have worked out some of our predispositions in the realm of money based on our specifically human neural circuitry, and their potential relevance to climate change.

Unexpected Financial Reward

Your reward system is not designed just to tell you if something is absolutely rewarding or not—but rather whether the reward you get is more than you were expecting. So, with the reward of money, if you get twenty dollars but were expecting ten, fMRI shows that this activates your midbrain dopamine output center, with its connections to that bland-looking but powerful data integrator, the nucleus accumbens.[10] This functions at the intersection of a number of scintillating neural highways: your memory circuits; your input from life experience and judgment from your various prefrontal and anterior cingulate decision-making circuits; your amygdala for various negative and positive approach and avoidance experiences; and the action of your motor circuits that let you reach, push a button, adjust your body language, and communicate with gesture and verbal output—in this case, responding to an unexpected reward. To illustrate this coordination of neural circuits, picture the elated facial expression, the exultant voice, the explosive air punch of the person who just got the match, shouting "Bingo!"

Conversely, if you get the same amount—twenty dollars—but were expecting thirty, something entirely different happens. Your VTA neurons do not release dopamine, and the dopamine level in the nucleus accumbens (in this case, measured in rats) actually decreases.[11] In addition to cells specifically tuned to respond to unexpected reward, we also have cells tuned to unexpected disappointment. We seem to be built with considerably more of both positive and negative "surprise" neurons than with neurons that just respond to an expected reward.[12] This design favors learning new things to adapt as your circumstances change; having the equipment that's predisposed toward strengthening these firing patterns means that next time, you'll be influenced by surprise consequences to be more inclined to do—or to avoid doing—that same thing again. Like the ancestor who figured out that berries tend to occur in association with specific features of the landscape, you might have stumbled on a great bargain on the third shelf in the corner of your favorite store—and so now you look there every time. You might even check out a similar location in a completely different store—all without necessarily thinking about it consciously. This reflects your brain, doing its thing to maximize your chances of staying alive.

As we saw in Chapter 2, for humans in specific circumstances involving money, dopamine release seems to be the product of two processes—how much you make after a given choice, modified by how much you might have made, if the circumstances had been different. If you gain money but might have gained even more, the dopamine release associated with that gain is lessened. Conversely, if you lose but might have lost big, the decrease in dopamine is moderated. Thus, in people, reward appears to be computed by your brain not just in absolute terms—that is, what you actually got or lost—but an additional computation is made based on the knowledge of what you could or should have gotten. Inputs from parts of the network we have encountered previously appear to serve as the neural correlates of "regret" and "relief" as they influence "how good" or "how bad" someone feels after learning the outcome of an economic decision.[13]

Your Brain Compares Your Financial Rewards to Others

Your nervous system's design has other features for giving you an advantage. With money, the reward value perceived by the brain is influenced not just by comparison to what we expected or might have gotten, but also by how we compare to others around us. The same twenty dollars elicits more of a reward response on fMRI if another person playing the same game and solving the same problem correctly only got fifteen.[14] In fact, getting the correct answer when someone else playing with you does not (whom you don't see and don't know, to eliminate any personal "sympathy" factor—more on this later) also is associated with increased activation of your nucleus accumbens. This phenomenon has been called "the joy of winning," but the smarty-pants personalities among us might also call it "the joy of being right."[15] While one might ascribe this to "human nature" or "common sense," similar trends have been shown in nonhuman primates as well.[16] Rather than chalking this up to some shrug-of-the-shoulder "human nature," this predisposition can be seen in a different light—it arises from the way our nervous system was shaped by evolution to process reward. Getting the right answer, getting a lucky break, and getting more than you expect appear to be, on average, even more rewarding when people can see that their situation compares favorably to that of a stranger in similar circumstances.

That the brain operates in this way may not seem surprising when sized up against what we all experience in everyday life. I know many physicians who make a very comfortable living but are dissatisfied, because they know a college classmate—one who didn't even study very hard!—who made it *really* big on Wall Street. This tendency may help explain the common phenomenon of escalation of expectations. We are happy for a short time with a particular reward, in the transient way that our reward system is designed to work. But when we recognize that we could have, or others do have, even greater rewards, our satisfaction becomes neutral or even negative to the rewards we do get if we don't get what we think is possible or get as much as others do. So, we humans, like the species from which we evolved, are finely tuned to register and learn from not just the value of reward, but whether we could be doing better if we tried something different. Like rats, our brain is designed for learning—for making associations that signal more reward than expected, or less. It accomplishes this by an extraordinary system designed to immediately evaluate the results of each behavioral choice, compare it to memories and perceptions of what has happened in the past and what is potentially available in the individual's observed world, and store the outcome using multiple networks calling in vast stores of information processed with myriad minuscule intersecting neurochemical signals.[17] Even the most trivial decision likely involves millions of synaptic interchanges involving networks spread all over your brain. In humans, money acts very similarly to food when we are hungry and water when we are thirsty; the same basic networks are involved. While scientists have unraveled details of specialization, such as specific sites within the network in which tangible versus abstract rewards are processed, these integrative processes are constantly on the lookout for something that may give us an advantage in the future.[18] If we are aware of enhanced possibilities, the function of our neural equipment is biased by the way our dopaminergic reward currency is calculated and dispensed, to nudge us toward behaviors that seek more, more, more. We were proud and satisfied with what we got paid for raking the neighbor's lawn until we found out that the kid down the street made more for raking the smaller yard next door.

Prosocial Rewards

While the tendencies outlined here likely jive with most people's life experience, it's clear that this can't be all there is to it. It's obvious that despite these tendencies, people don't always behave selfishly, trying to get more and more all the time—their behavior is tempered by many other factors. In fact, people are often generous, unselfish, kind, and heroic. We are able to both compete and cooperate. The lens of neuroscience can be used to look at these traits as well, because these too depend on how our brain evolved to work, to help us to survive, individually and collectively.

As a species we needed to cooperate. This is because on average, we were more successful at survival with a distribution of talents and skills than as solo practitioners trying to do everything ourselves, each as an individual.[19] Some people are good at mechanical skills, some at storytelling, some at problem solving, some at diplomacy, some at aggression, and some at comforting. Another essential trait we require is generosity—if we couldn't share, we'd be in big trouble, even before kindergarten was invented. Some people have this ability more than others, but as a whole, groups need some generosity to survive. Cooperation to distribute tasks, jurisdictions, and resources requires compromise, sometimes putting the needs of the group at large or other individuals ahead of our own immediate needs or preferences. While much has been made of unselfish behavior toward genetically related individuals being "wired in" in order to "protect" our own genes, people and other animals also act unselfishly toward unrelated individuals.[20]

Our working premise in this discussion is that *we simply wouldn't exhibit these kinds of behaviors if there weren't some kind of reward involved.* During the millions of years of evolution, our brains developed mechanisms to include reward for unselfish behavior. Like all traits—strength, empathy, endurance, numerical prowess, cheerfulness, rule-following, creativity—some people exhibit this tendency more than others.[21] Scientists have studied these types of behaviors using different labels—prosocial behavior, egalitarianism, generosity, inequity aversion, cooperation. That some kind of neural mechanism must make such behavior worthwhile to the individual seems obvious, but how can you study it? To get at the brain functions responsible, versions of the Dictator Game have been used to study the neural underpinnings of generosity, and the Prisoner's Dilemma to study cooperation.

One way to study human traits like generosity is to use games in which resource sharing between people is manipulated, most often again using money as a conveniently quantified resource. As one example, researchers investigated brain activity in specific locations using fMRI when money was distributed between two people in different ways. They wanted to know whether reward circuitry was activated when monetary inequality was remedied. If so, was it only rewarding for the person who had less to begin with and then got more?[22] In this experiment, two players each got thirty dollars. One of the two then received a "bonus" of fifty dollars, while the other player got no bonus. Next, while in the scanner, players reacted to additional money distribution combinations, including one in which the player who got the bonus got even more money, and others in which the other player got more instead, making the distribution more equal. The researchers found that on average, subjects' reward circuitry (nucleus accumbens and prefrontal cortex) was activated when the distribution was more equitable, even when this occurred because the other person received money that brought the two players into balance. While this result might seem to be in direct contradiction to data we've seen previously in which people generally respond negatively when other people get more than they do, note that this experiment isolates a very specific circumstance—restitution of chance inequality in the situation of a windfall. In this experiment, the money wasn't "earned" by performing a task—it was just handed out. This evidence suggests that among all the microscopic reward circuit components weighing in second to second, one inherent type of circuit responds positively toward equity, and against inequity.

In another example, after taking a medication that increases their dopamine levels, subjects playing the Dictator Game are more likely to give an equal amount of money to someone else as they keep for themselves (in other words, behave with more egalitarianism). This suggests that a reward, mediated by dopamine, is involved in encouraging individuals to behave fairly and share resources.[23]

With that in mind, when monetary gain and cooperation are at odds, which one wins? The answer is—as with so many things in neurobiology—it depends. Your nervous system "decides" what matters most at any given time and circumstance, based on your specific state and situation. For almost every question and variable you can imagine, researchers somewhere have designed experiments to study that particular circumstance. Results are not always consistent, but often show trends in how, on average, the nervous system typically works when facing decisions about money and other resources. Are you flush, or in debt? Independent, or responsible for others? Feeling secure, or smarting from a setback? Feeling connected to others, or alienated?[24] Under a greater or lesser influence of particular hormones or neurochemicals? Stressed, or calm? These are all words that describe influences we recognize experientially that are translated by neurobiology into minuscule, ever-changing counterweights that open and close the gates of our neural pathways leading to individual decisions. The reward system is the integrator that acts as the scale, summing up all the tiny weights and sending the outcome on to the basal ganglia to be translated into action. But as a general rule, most people inherited behavioral tendencies that would have promoted survival in times of scarce resources and when considerable energies had to be expended to attain them.

Agency

There are other experiences that have been shown in humans and some (other) animals to be associated with reward. One of these has been termed "agency"—the sense of accomplishment one gets from solving a problem, completing a task, "making something happen."[25] Both animals and humans get a reward *just from being able to make something happen.*[26] This is particularly obvious when seen in association with new learning—exactly the job of the reward system. It's common to witness a baby engrossed in opening and closing cupboard doors, expressing glee when getting the right shape into the right opening in a shape-sorting box, and trying over and over until figuring out how to toddle. Who isn't satisfied by mastering a new skill, getting better at a hobby, coming up with a solution to a vexing problem? Indeed, researchers have seen these same traits in animals from crows to monkeys. It isn't hard to understand why our nervous systems would have evolved to make these things rewarding—they are essential to learning, problem-solving, and adaptation. The reward aspect feeds directly into the "motivation" aspect that encourages the individual to persist at the task.[27] In fact, brain activation seen on fMRI has shown that people who are particularly motivated to be "at the top of the class" are more rewarded when a task is perceived as difficult.[28] When social scientists and political philosophers talk about satisfaction versus alienation in the work world, the discussion often centers on the sense of agency and accomplishment; to deprive people of this obvious reward is thought by some to be tantamount to making a job "soulless"—perhaps another way to express the sense that such work is dissociated from what our brains evolved to find rewarding. It was agency, combined with his dedication to saving his son's life, that fueled John Holter's invention we encountered in the Introduction.

History provides vivid examples of how the rewards of money, competition, risk-taking, and agency, and their particular strength in certain groups of predisposed individuals, smoothed the way for climate change. In her 1904 book *The History of the Standard Oil Company,* journalist Ida Tarbell paints for her reader a colorful portrait of the potent rewards of agency and profit to be had in the fledgling oil industry in the late 1800s.[29] The cast of characters embodies the diversity of neural traits we saw in Chapter 2, particularly attracting those with an affinity to high profit, a confidence in their own agency, and an eagerness and predisposition to gamble. The traits in these early adventurers mirror that subset of Parkinson's patients who were found to be at highest risk of developing gambling addiction, when dopamine as medicine was added on top of their existing predispositions. Tarbell writes,

> On every rocky farm, in every poor settlement of the region, was some man whose ear was attuned to Fortune's call, and who had the daring and the energy to risk everything he possessed in an oil lease. It was well that he acted at once; for, as the news of the discovery of oil reached the open, the farms and towns of Ohio, New York, and Pennsylvania poured out a stream of ambitious and vigorous youths, eager to seize what might be there for them, while from the East came men with money and business experience, who formed great stock companies, took up lands in parcels of thousands of acres, and put down wells along every rocky run and creek, as well as over the steep hills. . . .
>
> As men and means were found to put down wells, to devise and build tanks and boats and pipes and railroads for handling the oil, to adapt and improve processes for manufacturing, so men were found from the beginning of the oil business to wrestle with every problem raised. They came in shoals, young, vigorous, resourceful, indifferent to difficulties, greedy for a chance, and with each year they forced more light and wealth from the new product.

Novelty

That change and new stimuli are rewarding has long been understood by psychologists, child developmental specialists, writers, and philosophers. In recent years, the ability to link novelty to reward biology directly has led to new insights into the mechanisms of this association and how it may vary among individuals.

It is worth noting that in simple species such as crayfish, who sport all the basic reward system ingredients we have described in previous sections, the reward of "novelty" is actualized as egging behavior toward increased time spent exploring their surroundings.[30] It is not difficult to see how this could be useful to an animal—exploration increases the chances of finding something tasty, as long as mechanisms for sensing danger coevolved, so you don't explore yourself right into the path of a predator. Mammals from rodents to pigs to monkeys pay more attention to novel objects than to familiar ones, as do humans from infants to adults. It makes sense that attraction to novelty is critical to learning during the early years of childhood; many studies in the education and psychology realms have shown that even very young babies look preferentially and longer at things they haven't seen before.[31] Common experience bears this out; from the new toy to the new kid to the new entertainment release, "new" generally is perceived at first pass as a potentially positive thing.

How does brain design nudge you to treat new as attractive, so that you're interested in learning something that might be helpful? Memory circuitry is important, as a new stimulus must be compared at lightning speed to what you've experienced before in order to recognize its newness. The brain is remarkably good at this, in virtually all kinds of animals, including humans—it's a very old, ingrained ability.[32] Once recognized as new, novel stimuli have been shown to correlate with the release of dopamine in the reward circuitry, quickly followed by another signal depending on whether the novel stimulus is actually rewarding or aversive.[33]

The positive reinforcement of paying attention to something new has been termed an "alerting response." Specific neurons perform this crucial task, and integrate it with other parts of the system to translate all of these dopamine nudges into actual behavior and learning, worked out cell by cell in animal studies. Interestingly, because of feedback loops and complex signaling pathways involving dozens of specialized nuclei, the system is designed with all kinds of possibilities in mind. For example, reaction to a novel signal is suppressed if attention is concentrated on another life-critical task at the same time.[34] Thus, while novelty is rewarding in general, our brains are sufficiently flexible that we won't be distracted by something novel at the expense of getting an immediate task done that is crucial for survival. This too is a learned process, and we generally are better at it with practice and a richer trove of experiential data with which to compare the novel intrusion. Our reward system is designed beautifully to facilitate that learning—and to enhance survival in a world full of things we've never seen before, things that remind us of other things we've seen, and things we know well and have already learned, from experience, whether they're good or bad.

Like other traits, attraction to novelty varies among individuals; any parent or teacher can attest that some kids dive headfirst into anything new and perceive it as exciting with little forethought or regard for consequences, while others tend to hold back and exhibit caution. One of the keys to human survival as a species was the distribution of different traits among individuals; each has an important role to play in an optimally functioning and adaptable society.[35] If all members of the tribe ate a new but poisonous food, there goes that branch of society; a few cautious holdouts would save the group from extinction. Likewise, when resources in one place became perilously scarce, the adventurous ones who were willing to journey to new places—and succeeded in finding new food sources or better protection—would keep that lineage alive. So both tendencies have their use in avoiding extinction.

Like many other traits, attraction to novelty can be measured using behavioral checklists. Patients with Parkinson's disease who score high on tests of attraction to novelty are more likely to develop problems with pathologic gambling from medications or surgical procedures that "rev up" the dopamine system; presumably, these patients were starting at a higher baseline, and the extra boost tipped them into pathological overdrive.[36] Indeed, healthy individuals with specific subtypes of dopamine receptors on their neurons are more likely than others to be attracted to novelty, reflecting the potential role of genetics in many behavioral traits.[37] Neuroscientists have identified high and low "sensation seekers," with those found to be "high" especially drawn to novelty but also having an increased tendency to engage in risky behaviors—the "natural" entrepreneurs, and in some cases, the future addicts and criminals. The differences in activation patterns in brain structures we met in Chapter 2 that are involved in control and monitoring processes, reward circuitry, and translating reward into behavior can be seen in such individuals throughout the lifespan.[38]

The reward of novelty has an obvious association with consumption. Your new living room furniture makes you feel terrific every time you see it—for a few months. Then the redecorating urge resurfaces, and you want a new carpet to make it look even better. Brain rewards are designed to be transient, so that you learn new things in new circumstances, and advertisers and marketers capitalize on this to convince you that you need something new to "complete the look"—but, of course, it's never complete. The advertisers and their furniture store employees' rewards come in the form of a job well done, profits increased, praise, publicity for the company, doing better than their competition, a year-end bonus. Their rewards in the context of work rarely come from doing something to create a healthier planet.

Familiarity

But wait, you may say—sometimes I just want things to be the same. I want to know that my coffee will be the way I like it in the morning, I want to drink it from my favorite cup, and I want my comfy, familiar old bathrobe after my shower. I am sentimentally attached to an iconic building in my neighborhood that's been part of the landscape since my grandparents were children, and I signed a petition to try to save it. We value family traditions, and I read my kids the same books—in fact, they themselves insist on reading the same books over and over.

So if our reward system is designed for novelty, why do we derive comfort from things staying the same?

The neurobiology of familiarity has been less studied than has that of novelty, but some experiments shed light on the complexity of behavioral response to these seemingly conflicting preferences. Our brains are designed for many circumstances, with amazing built-in flexibility. In rats, stress (caused by a period of confinement in a narrow tube) causes preference to shift from novel to familiar objects, and this result can be replicated by giving the rats a dose of the stress steroid corticosterone.[39] This makes sense from a survival point of view—when conditions are not optimal, sticking to the familiar may be a safer bet.

There are other biologic and, obviously, cultural influences. Many studies have investigated differences among people that might shed light on the tendency to prefer novelty to familiarity. This research has been generated from a wide range of disciplines, including political science, economics, psychology, and neuroscience.

A tendency to prefer familiar things increases with age, with children tending to respond more strongly to novelty, while adults respond less strongly and tend to exhibit a preference for familiarity.[40] These observations align with the fact that the driving developmental goal of children is to learn enormous amounts of material, and attention to novelty is a crucial part of the learning process for both visual and auditory stimuli, including language.[41]

The Formation of Preferences

Preference for familiarity also has been shown to correlate with personality traits characterized by a tendency to worry about negative things happening rather than feeling positive about possible rewards in something new.[42] Along similar lines, in the broad context of social values and political behavior, researchers generally define "liberals" as having the tendency to be more open to new experiences and viewpoints and exhibiting tolerance for uncertainty and ambiguity, whereas "conservatives" generally value stability and tradition and show more structure and persistence in their approach to decision making.

What aspects of brain function determine these kinds of differences? When talking about behavioral tendencies, it's important to understand the distinction between the terms "genetic" and "biological." To say a difference in a behavior is "genetic" is to suggest that it is related directly to an inherited trait—which, of course, can mean a difference in how the brain is predisposed to work. An example might be an inherited predominance of a particular type of receptor for a specific neurotransmitter, which might make it work faster; this might tend to make someone behave in a particular way under certain circumstances compared to people without so many of those types of receptors. In contrast, the brain can be altered by experience. Thus, to say that a behavioral tendency is determined by environment or culture is not to say that it isn't "biological"—it just means that the brain can be altered by the environment in a way that isn't dependent on the specific genetic profile of that person. Of course, these factors can interact. Having a particular gene may make individuals more likely to be influenced by their environments and experiences in a particular way—what researchers call the "genotype-environment interaction."[43] And to add yet another layer of complexity to the nature / nurture interaction is the burgeoning study of epigenetics—environmental events that affect how genes are expressed, typically via chemical processes that influence how the genetic instructions are "read" and thus can influence traits and behaviors exhibited in subsequent generations.

With respect to sameness and novelty, the related tendency to be more conservative versus liberal in one's views has been studied in large numbers of twins and their parents. Identical twins share all the same genes, while they share only half their genes with each parent. To take advantage of the opportunity provided by nature to understand the role of genetics versus environment, large databases of questionnaires have been amassed from tens of thousands of twins and their families. If twins are more like one another than they are like their parents or non-twin siblings in a particular trait, a genetic basis for that trait is supported. In contrast, traits that correlate more with a shared environment than with shared genes suggest that your upbringing or culture is more likely to be responsible for your behaviors.

Perhaps not surprisingly, different experiments have found that being "liberal" versus "conservative" can be explained by both genetic and environmental differences.[44] While the specific brain functional correlates remain incompletely understood for such diffuse processes, researchers have found some evidence for differences in brain processing in reward network structures and subtypes of neurotransmitters that are inherent—and inherited—that might explain some of these differences. For instance, increased activation of the anterior cingulate cortex, the structure implicated in decision making under conditions requiring cognitive control (Figure 2), has been correlated with liberal views, while a greater number of dopamine receptors favoring an enhanced fear response to threat may correlate with the "negativity bias" attributed to conservative cognitive styles—the tendency to resist change as something that might be threatening or harmful.[45] These differences may have implications for attitudes toward environmental issues, as we will see when we look into pro-environmental behavior change in Part 3.[46]

Another aspect of brain function that might help explain the yin and yang of novelty versus familiarity is the study of how preferences are developed in the first place. This has been the object of study by diverse disciplines, from agriculture to advertising. To see how preferences form, let's start our science probe on . . . the farm.

Agricultural researchers want to know how to make livestock farm animals grow faster by encouraging young animals to eat more, thus increasing the farmer's yield. Since feeding choice, like most preferential behavior, involves the reward system, scientists wanted to understand how brain networks might be involved during the time these preferences are learned early in life. In a study of feeding preferences in piglets during weaning, most young animals initially preferred feeds flavored with sweet citrus extract over plain wheat-based feeds. However, once they had been fed either plain or flavored feeds for several weeks, they tended to choose the feed with which they were most familiar. To understand what was going on in the animals' brains, anesthetized piglets were exposed to the taste and smell of the different kinds of feeds while their regional brain metabolism was measured by PET scan. Piglets accustomed to the flavored feeds activated more reward-related circuitry when exposed to the taste and smell of that type of feed than did piglets raised on plain feeds. So, while they might choose the feed they are used to when given a choice, those accustomed to the flavored feed got an "extra boost" of reward activation. This kind of mechanism might explain why it's hard to go back to simpler, plainer things once you've acclimated to "the good stuff."[47] You can learn to like low-fat milk in your coffee if you're used to cream, but it takes a long time to get over that high-calorie reward value. A spare, uncluttered home eventually feels calming, but for a long time the "stuff" may be missed. Common experience is full of the letdown of reward after a period of athletic prowess or intense career success when life reverts to the more mundane. Simplification takes time, effort, and practice before alternative rewards kick in. No wonder people take the easier route! This tendency to acclimate to "more," built into our neurobiological inheritance, makes it challenging for us to do with less in all spheres and likely contributes to accelerating consumption—not just of goods, but of all the activities that support and drive our growth-based economic engine and also accelerate climate change.

A totally different realm in which novelty versus familiarity has been studied involves interpersonal relationships. If novelty calls the shots, why do marriages or other long partnerships ever last? One line of evidence from the brain is that hormones also play a role in the reward of novelty versus familiarity. Oxytocin seems to bind family members together, and when increased through activities that cause it to rise (such as physical contact) or by pharmacologic administration as part of an experiment, it increases the strength of the reward one gets from looking at pictures of long-term mates.[48] Evolutionary pressures tweaked reward values by adding other biological modulators that promoted survival of the species. Hormones are powerful modulators of how rewarding something is, just as hunger enhances the reward of food-related stimuli.[49] Similarly, fMRI has documented the involvement of reward circuitry, empathy, and stress buffering in happy long-term marital relationships.[50]

Can things be both familiar *and* novel, appealing to us through both kinds of reward? Of course—and in fact, that's probably usually the case. There are a million common examples from everyday life. You like someone's house because it reminds you of your favorite aunt's house. You like a color because it played a role in a happy time in your past. A new person has a similar sense of humor to someone you valued in a previous period in your life. Novel, but familiar. There are individual differences and circumstantial changes—hormones, stress—that may tilt us toward finding novelty versus familiarity more or less rewarding at any given moment. Our brains are designed to make these kinds of associations to help us recognize patterns that optimize successfully navigating new needs and changing circumstances.

Cultural Learning and Reward Valuation

As we have seen, some tendencies have been shown to cross cultures, such as the "irrational" choices people make about money, but it is also obvious to any observer that people vary individually and as members of particular groups. The fact that one culture values shell beads and another prizes high-performance automobiles may just reflect availability. But how do we *learn* what's valuable? The obvious answer is that we learn it from the people from whom we learn most things—our elders. Later in life, we learn from our peers, and studies have tracked at which ages and in which societies these influences shift.[51] This is likely true for assigning value both to things and to behaviors—what can be called "cultural values."

One can make a nearly endless list of values that vary among cultures and argue about their operational definitions—honesty, loyalty, bravery, skill, stoicism, sociality, nurturing, and so on. In one culture, the person with the fanciest clothes, biggest and most ornate dwelling, and most extensive property is accorded the most respect and deference. In another culture, behaving in such a way that your assets are obvious is considered crude and is cause for denigration and even ostracism. How could two such opposite behaviors be valued so differently?

As an illustrative example of cultural values varying with context, consider the case of cats. Imagine this scene: You and your family, including young children, go to the center of town for a family-friendly outing. You're going to what is one of your community's recurring big public events—something everyone attends and anticipates eagerly. As you approach the center of town, you hear screeches and howling, getting louder and louder. Your kids, knowing what this is from prior experience, laugh and clap and run ahead toward the noise with happy exuberance. There are plumes of smoke and hundreds of people craning their necks to see. Everyone is cheering and shouting out their approval.

As you work your way through the crowd, keeping your children just within your sight, you can catch a glimpse of the main event. Over a roaring fire, cats are suspended in baskets or by their tails, being burned alive. The more they scream and struggle, the happier the crowd is, and the louder their roars of approval.

Are you horrified? How could something so unequivocally cruel and awful—the public torture of sentient, live animals—be the source of merriment and approval?

Such events occurred regularly in Europe during the Middle Ages. People were taught by religious leaders, who were significant cultural authority figures, that cats were the agents of the devil. Torturing cats was considered equivalent to torturing Satan, and so was considered worthy, just, and good. The more agony the cats exhibited, the more Satan was being denigrated and subdued.[52] What a show!

Contrast this medieval European phenomenon of cat denigration and torture, elements of which extended into the eighteenth and nineteenth centuries (and, incidentally, contributed to the Black Plague, which was spread by rats), with the attitudes toward cats in ancient Egypt. There cats were venerated, associated with an important goddess, and treated as valued and wise members of the household. That their useful role in controlling rodents and snakes may have influenced human attitudes is likely, but this was translated into cultural and religious values taught by societal authorities.[53]

Some authors have pointed out the evolutionarily adaptive tendency of humans *to believe what they are told by authority figures.* This is adaptive, because if you didn't stop when your parent or grandparent or elder shouted "STOP!" when you were about to run over the edge of a cliff or poke at a beehive or step on a snake or fall into a river, you would not have survived. If you didn't learn that eating this but not that was safe and that behaving in a certain way was acceptable in society but behaving in another way could get you ostracized and fending for yourself, you would be in a much more precarious situation. Thus, tending to believe what we are told by those in authority—parents, elders, and societal leaders—has had survival value, and so is a powerful part of the way our brains learn. While every society has people who are more or less inclined to believe rather than question authority, as a general rule, people have a tendency to be *gullible,* defined from a neurological point of view as a general willingness to believe and to imitate what is taught by those in positions of authority.[54] This is particularly, but by no means exclusively, a trait exhibited early in childhood.[55] What people are taught early in life has particular tenacity, especially when coupled with the idea that deviation will lead to punishment or even eternal suffering.[56] But this tendency is seen also in adults. In my field of surgery, what you are taught by your mentors during training is so difficult to override that some surgeons never can accept new methods for performing an operation.

Thus in many contexts and life stages, people come to their opinions about what is important, valuable, and in many instances rewarding based on cultural learning. This learning does not occur in isolation from the circumstances, society, and beliefs that are part of the experiences and culture of each individual, and can be quite arbitrary and readily influenced by chance factors.[57]

How does all this swirling neurobiology relate to consumption and other behaviors with environmental relevance? We've seen how we are predisposed to be both competitive and cooperative. We're generally attracted to novelty, as well as to familiarity; we are influenced by our inherent biology and shaped by our specific experiences and cultural context. While preferences and beliefs learned early in life may be strongly ingrained, the brain is sufficiently adaptable that these can change, both in the life of an individual, and in a society.

Thus, the human brain, built on eons of engineering designed for short-term survival, has predispositions inherent in its structure and how it interacts with the world to function. However, it also is designed to be adaptable to new circumstances—but can change most easily under particular types of influences, all of which require the involvement of the reward system.

The circumstances in which we find ourselves today differ from those during most of our evolutionary history. The Anthropocene, only the last part of the last step of our evolutionary timeline from Chapter 1, presents entirely new kinds of survival challenges than our brains have faced before. Our neurobiological heritage helps explain why we find ourselves in our environmental dilemma. This knowledge, while important background, by itself doesn't provide a path forward; we need to continue our exploration to learn what might help us to emerge.

If behavioral choices guided by human tendencies regarding money, agency, and other rewards got us into this fix, is there any hope of changing our behavior to escape the worst-case environmental scenarios? To answer, we need to better understand which specific behaviors need to change, and whether there's hope for making alternative behaviors rewarding. We need to place our behavioral choices in the context of what we can learn from environmental scientists, as well as the science of behavior change. Can we substitute less destructive options, or enhance the power of social rewards and equity to make alternate decisions that minimize environmental damage, at a scale that could make a difference? Our lives and our societies continue to change at a rapid pace, and that acceleration figures into the equation as well. How these changing circumstances intersect with human reward, behavior, and the environment will be explored as we continue our journey. We need first to answer a question about one additional reward for human beings that we haven't addressed so far: To what extent is nature itself rewarding to people?

4

Biophilia and the Brain

WHEN I WAS A RESIDENT AND, later, a staff neurosurgeon at a children's hospital in Philadelphia, as soon as the weather started to turn warm in the spring, hundreds of hospital employees could be found outside at lunchtime. The concrete benches behind the hospital buildings were rarely used; instead, everyone piled out of the glass doors of every building in the complex, crossed the busy street, and occupied every available inch of grass on the steeply sloped lawn of the museum across from the hospital. It wasn't a big space, and not the most convenient; it was tilted, there were a few scattered trees, and the view was of the street—nothing special. But there was something about sitting on the grass that just seemed to draw people; we called it "beaching it." Two decades later, the same thing happens on the Bulfinch Lawn at my hospital in Boston, an acre or so of sloping grass in between the disparate clutch of buildings that have risen up over the centuries, old and stately anchoring new and glassy. While this land is extremely valuable real estate and could be used to expand—a near-continuous pressure on most academic hospitals—nothing has infringed on the Lawn. People don't necessarily think of why this is. It just seems obvious, like a sacred space you wouldn't dream of disturbing.

We've now talked about rewards from food, from money, from novelty and familiarity, social rewards, problem solving, agency, and others. The broad topic at hand is what it is about the way our brains work—in particular, how we are designed by evolution to process and feel reward—that has allowed us to accelerate behaviors that run counter to what's needed to change our trajectory of environmental decline. Despite our collective behaviors to the contrary, some scientists and writers feel strongly that most people find the *natural world itself inherently rewarding.*

But if this is the case, why do we have so much trouble prioritizing the protection of the natural world? Alternatively, is our "love of nature" specific to certain people and circumstances, perhaps especially in modern times, and not a strong inherited trait at all? Maybe our inheritance is more likely to predispose us to fear nature or simply to interact with it only as much as necessary for survival. These tendencies will be important to unravel as we seek to understand the role our brain design plays in our current environmental crisis. In this chapter we will review evidence for and against the existence of biophilia; its forms, frequency, variability, and potency; and its implications for behavior change strategies that we will address in upcoming chapters.

A Brief History of Biophilia

The term "biophilia"—literally, love of living things—was first coined by psychologist Erich Fromm in 1964 to refer to what he conceived as an innate tendency of humans to be drawn to life forms found in nature.[1] Evolutionary biologist E. O. Wilson further developed and popularized this idea in his book titled *Biophilia* by characterizing this tendency as a genetically inherited deep drawing of humans toward the natural world and living things.[2] While philosophers and behavioral and social scientists have long discussed interactions among human beings, the idea of biophilia is that humans are also drawn to and comforted by being in nature, in the presence of vegetation, natural landscapes, and other animals in addition to other humans. Here we'll follow physical and behavioral health and architecture fields in the use of the term "biophilia" to mean a wide swath of nature-generated, as opposed to "built," environments and their components.

As with other human traits, it would seem from common experience that some people love nature more than others do. Is this because of an inherited tendency, because of early life exposure, or some other factor? Do we "imprint" on nature, like the baby geese that were convinced that psychologist Konrad Lorenz was their mother, because they were exposed to him at just the right phase of life? If that's the case, people in the developed urban world who have less exposure to relatively "untouched" expanses of nature just may not value it as much.[3] E. O. Wilson grew up exploring the woods and swamps of Alabama and Florida, master of small universes containing whole societies of vertebrate and insect inhabitants and divested of the human turmoil endemic in other aspects of life. This was a formative experience he has described as critically influential in his scientific views. But Erich Fromm grew up in urban Frankfurt, Germany, in a family that valued book-based scholarly work and Talmudic studies. His father ran a wine shop, and apart from occasional excursions to countryside resorts, Fromm's daily exposure to nature in early childhood seems likely to have been limited. People who have grown up in cities also may love nature, so it doesn't seem to be just about early exposure.

That biophilia is an inherent tendency that humans embody as part of their biology is a theory, or hypothesis, that is difficult to test directly.[4] Most of the evidence is indirect and deductive, but in aggregate, it has been accepted by many as a compelling case. Let's look at some of these arguments, because if there is evidence that nature is rewarding or beneficial, this may be relevant to the generation of environmentally proactive behavior.

One line of evidence for a universal attraction to nature that has been cited to suggest a shared biology is *landscape preference* among people from different places and cultures. A number of widely cited studies show that many people in disparate locations choose the same images as most appealing. First, by a wide margin, people rank landscape images as more attractive than those of buildings.[5] Second, among landscapes, specific features rise repeatedly to the top of the preference list.[6] These include an expansive perspective, as if the viewer were looking out from a hill or outcropping. The ground is green, covered with low grasses or similar vegetation, and there are scattered trees with some low branches and wide canopies. There is water in the picture, and maybe some animals.

Does that sound attractive? This is the so-called "savannah landscape," and some scientists hypothesize that humans perceive these features as positive because we learned evolutionarily to value these things for our survival. The elevated vantage point allows us to see predators or enemies; the vegetation suggests accessibility and potential food sources; the trees with low branches facilitate climbing for protection and canopies for shelter; water and animals signal opportunities for sustenance.

The observation that many people choose similar scenes as most appealing suggests to some authors that these preferences are inherited as part of our common ancestry, during which early humans evolved over millions of years in the savannah environment of East Africa. Ulrich and others expanded this idea to suggest that specific features of a landscape—an optimal amount of complexity, a sense that the area can be walked through and explored, depth and perspective suggesting a bit of mystery and potential beyond the next curve, and low threat—are preferred on a psycho-evolutionary basis favoring landscapes that embody the conditions optimal for sustenance and safety.[7] These features are the ones found in many parks as well as rural views considered scenic that encompass meadows, fields, and edges of forests. Landscape architects have used these data to design recreational settings, building grounds, college campuses, and roadside vistas to make them most attractive to people. The fact that real estate prices go up when the location includes beautiful views, gardens, coast, or lakefront provides market evidence that these things have value to humans for which they are willing to expend resources.

Another body of evidence used to support the hypothesis of an inherited tendency to be drawn toward nature is its opposite, so-called *biophobia*. During most of human evolution, the nature amid which people lived may have represented at least as much of a threat as a source of reward. Researchers and cultural historians have documented a nearly universal tendency to aversion to snakes and spiders; even adults, as well as young children who have never been exposed, tend to recoil from these squiggly living forms as their first reaction. This response to snakes is exhibited by other animals as well, including dogs and some primates.[8] Humans tend to instantaneously recoil much more from snakes and spiders than from those objects that are newer in human history but on average are far more dangerous, such as guns, frayed electrical wires, and motorcycles.[9] Snake and spider phobias are among the most common experienced by people, and studies have shown that twins typically share specific phobias more frequently than do more distantly related people in similar environments, further supporting a genetic basis for these fears.[10] Together these data are used to support the argument that biophobia represents an inherited neural wiring diagram that evolved to perceive risk in specific stimuli; therefore, say these theorists, it follows that attraction to the positive aspects of nature also is likely to be inherited.

Another line of inquiry comes from the disciplines of education and child development. A wide variety of studies have shown that children learn better when exposed to nature and the outdoors. They are less stressed and more creative, exhibit better attention and concentration, are more active, get along better with others, and are healthier.[11] For instance, Scandinavian kindergarten children who spend outdoor play time every day in natural and diverse landscapes of forest, shrubs, trees, rocks, and fields have improved motor and cognitive skills compared to children who use conventional flat playgrounds with manufactured play equipment. Children playing in outdoor natural spaces tend to find "functions" in the natural landscape; bushes become forts and palaces, rocks become pirate ships, cliffs become climbing and jumping places, often with very specific names invented and assigned by the children themselves—Eagle's Nest, Princess Palace, and so forth.[12] In conventional playgrounds with flat ground surfaces and playground equipment, children with the most athletic prowess tend to become dominant in the social group, while in natural landscapes, children with the most creativity are looked to as the leaders.[13]

Many educators and psychologists conclude from these findings that children are "designed" to learn from nature, and that play in humans, as in animals, subserves the function of learning about the world and acquiring physical and social skills. Supporting evidence comes from observing play across cultures, including in contemporary hunger-gatherer societies.[14] Given a choice of where to play, most children will choose a natural environment outdoors (except, perhaps, if the alternative is video entertainment, but more on that in the next chapter). Outdoor play spaces are among the most memorable for adults recalling their childhoods, and outdoor sites typically make the list of favorite places for subjects of all ages.[15] This kind of play is just plain fun. Think of what you see most kids do when let loose outside: They roll down hills. They climb and jump off rocks. They hide behind bushes and chase each other around and shriek and laugh (and sometimes get hurt and cry—learning as they do). They build things and use sticks to stir soup made from leaves and rocks and water and create "secret hiding places" and make up story scenarios and play characters that they act out with other kids. Like E. O. Wilson in the swamp, they experience the reward of agency when they aren't told what to do by adults, instead figuring out what affordances are provided by the natural environment to create their own world. And they express clearly that they perceive nature as inherently beautiful.[16] *If it weren't rewarding, we wouldn't do it.* The argument for biophilia arises from the observation that when left to their own devices, children are drawn to and prefer natural environments in which to play, have a greater diversity of play types and experiences, and learn while having fun. Fun is the experiential embodiment of the reward system doing its job.

Nature, Attention, and Stress Reduction

If we are, in fact, "drawn" to nature, what it is that draws us? We will address different definitions of "nature" in Chapter 8, but for this discussion we will use the term broadly to refer to predominantly nature-generated, rather than built, surroundings. Most people who have a positive reaction to time spent in nature describe it as "peaceful," "calming," "refreshing," or "restorative." Some researchers hypothesize that these subjective positive reactions to natural settings occur because nature, in contrast to our typical work and school demands, draws our attention effortlessly to shifting elements in the landscape second to second. In this context, we are freed from the typical modern-life mental effort required to direct our attention for long stretches at a time to a specific work or learning task—something that these researchers posit was not much a part of evolutionary life, when success and survival instead required constant *shifts* in attention to the constantly varying natural world. As psychologist Steven Kaplan notes, "It is only in the modern world that the split between the important and the interesting has become extreme."[17] According to this theory, it is the mental fatigue experienced from prolonged sustained attention to one task that exposure to nature relieves, because our attention is drawn to a variety of changing stimuli in the environment at a pace and variety to which our brains are evolutionarily tuned. Nature provides an effortless "attention work break" as our inborn responsiveness to features in the natural world perceives them as inherently attractive and interesting.

Another source of data about the interaction of nature exposure and attention comes from the arena of attention-deficit / hyperactivity disorder (ADHD). Some authors note that children with attention and hyperactivity disorders often function perfectly well in the natural environment—their ADHD is in essence "cured" in nature but becomes apparent in the classroom, where prolonged sustained attention, typically to tasks not of one's own inclination, is required. Writer Richard Louv coined the term "nature deficit disorder" to highlight that frequently shifting attention—one of the hallmarks of ADHD—may be a matter of context. He argues that children with little exposure to nature suffer from a mismatch between their "natural" abilities—what is required to be healthy and to develop normally in the natural world—and how behavior is constrained in day-to-day structured activities occurring within the built environment, thus representing another disconnection between our evolved nervous system and modern life.[18] On average, children with ADHD show less inattention, impulsivity, and hyperactivity while in a natural setting compared to a human-built environment. Furthermore, impulsivity and inattention are reduced after exposure to nature compared to other activities.[19] Similar benefits from nature exposure may accrue even to students who do not have attentional problems. High school students who have views of nature from the classroom have improved test scores and graduation rates compared to those without, controlling for socioeconomic and other variables.[20] Increased aggression in urban environments correlates with measures of attentional fatigue and is lower in individuals who live close to neighborhood green spaces.[21] Thus, there seems to be something about exposure to nature that gives us a break from the "work" of non-nature-induced sustained attention and also enhances our subsequent ability to pay attention when we are called on to do so.

In contrast to those studying attention, other researchers have focused on the "restorative" power of nature as a consequence of its capacity for *stress reduction.*[22] The argument here is that exposure to a nonthreatening natural landscape correlates with physiological changes that counter those elicited by stressful circumstances. As one test of this hypothesis, investigators induced stress (by asking subjects to watch a movie about disturbing industrial accidents) and then measured the physiological changes induced by visualizing natural compared to urban landscapes. They found that stress measures were reduced by viewing the natural but not the urban landscape.[23] The restorative effects of nature can't be solely about attention, say these researchers, because snakes and spiders draw your attention "naturally," but they don't relieve your stress! There must be something else our brains do in response to the positive aspects of nature that are rewarding.

These two concepts are not mutually exclusive and may differ primarily in the tools one has available to measure complex neural processes like "attention," "arousal," and "stress." Tools like subjective rating scales to capture what people can verbalize they are feeling and experiencing, brain wave measurements, and heart rate changes provide some insights, and have been translated into theories of restoration as a reprieve from the "energy" required for nonvoluntary attentional tasks or physiological perturbation engendered by stress. In fact, there is some scientific support for the idea that some kinds of mental effort can be perceived as fatiguing and as stressful, and that this does indeed reflect brain "work." For instance, patients with mild memory impairments have subjective perceptions of having to "work harder" to perform mental tasks, and this is reflected in larger areas of cortex seen on fMRI that must be recruited to perform the task compared to normal controls. That is, people generally can in fact perceive when they have objective evidence of "needing to use more brain" and having to work harder to perform a required mental task, and they perceive such demands as both fatiguing and stressful.[24] But even fMRI is a crude measure of the trillions of events happening in your brain at any one moment, and we don't fully understand why exposure to nature makes people feel better. That said, many experiments measuring stress perception and attention performance in subjects exposed to natural environments compared to other types of environments have confirmed the beneficial effects of nature on these functions.[25]

A body of research on the power of affiliation between humans and animals also has been used to support the biophilia hypothesis. Well over half of US households own pets; this tendency of humans to keep animals in their lives, even when they engender added expense and work, has been used as evidence supporting the concept that humans are drawn to affiliate and bond with other living things. While direct interaction with animals is no longer necessary for survival for most people in industrialized societies, it is notable that people have not just walked away with indifference; animal interactions and interdependence in hunter-gatherer and agrarian daily life has been supplanted by birdfeeders, aquariums, zoos, and household pets.

And does interaction with animals have similar beneficial effects to those shown for being exposed to natural landscapes? Data are mixed. Owning a pet has been shown to improve behavior in adolescents with attentional and conduct disorders and to encourage children with autism to talk and to interact, even when other approaches have failed.[26] While a number of studies show the beneficial effects of pet ownership on socialization, exercise, longevity, and life satisfaction, many of these can be criticized on methodological grounds; unlike studies on the effect of a drug, you can't give a placebo animal to half the subjects and look at the effects on their stress levels and happiness.[27] Some studies have shown no (or even negative) effects of pet ownership on health and psychological factors like life satisfaction and loneliness.[28] However, the fact that so many people voluntarily own animals can't help but add some credence to the general idea that humans, for whatever reason or with whatever effect, will put forth effort and utilize resources to affiliate with animals even when they are not necessary for basic survival needs. *If it weren't rewarding in some way, we wouldn't do it.*

Evidence from Health Care

The practice of exposing patients to gardens and other forms of nature to aid recovery is almost as old as recorded history.[29] One of the first modern experiments addressing the health effects of nature exposure was published in the prestigious journal *Science* by environmental psychologist Roger Ulrich in 1984. A group of patients who underwent the same operation (gall bladder surgery) at the same hospital by the same doctors were studied to see how fast they recovered, how much pain medicine they required, and how quickly they were discharged from the hospital based on one variable: whether their hospital room looked out on a brick wall or on leafy trees. All other factors being equal, the tree-viewers required less pain medication and left the hospital on average one day sooner than the brick-wall patients.[30] These striking results were perhaps the first to gain widespread recognition within the medical establishment that nature exposure may help people to heal. Other research corroborated these findings. Patients undergoing thyroid surgery or appendectomy had shorter average recovery times, required less analgesic, and rated their experience better if their rooms contained a variety of potted plants and flowers compared to patients in rooms with no plants.[31] Patients recovering from cardiac or pulmonary disorders whose rooms had panoramic window views reported improved physical and mental health symptoms, albeit with some variability of the effect among patients.[32] These results have spawned a wave of incorporation of "healing gardens," horticulture programs, and other green space into the designs of acute care hospitals, mental health facilities, and rehabilitation centers.

Other investigators have studied the relationship between individual exposure to nature and health outcomes once patients leave the hospital. Among patients with stroke in a large metropolitan area, those who lived close to surrounding vegetation (as measured from satellite maps) on average had lower risk of mortality over the subsequent five years, even when adjusting for other risk factors.[33] A related study used a "natural experiment" in Michigan to look at the relationship between loss of trees due to the parasitic emerald ash borer and patient deaths due to cardiovascular or respiratory causes. The researchers found that as more trees died, the incidence of deaths in those neighborhoods increased. While the study couldn't determine the exact reason for the increase, the researchers hypothesized that both improved air quality and stress reduction provided by trees might play a role.[34] Thus the notion from ancient times that exposure to nature is beneficial to healing after illness or injury has been reinforced with considerable consistency by modern scientific inquiry.

Cross-cultural data suggest a similar incidence but poorer prognosis for schizophrenia and other major psychopathologies in developed countries compared to those with more traditional agrarian lifestyles. Additionally, surveys show an increased average sense of well-being in rural cultures compared to urban and industrialized settings. Certain types of psychopathologies are found at higher rates in modern "manufactured" lifestyles, including depression, suicidality, and drug abuse. These findings have been interpreted as evidence supporting the idea of biophilia and are hypothesized to reflect the consequences of increased isolation from others as well as separation in modern life from regular interaction with nature, to which we are generally better adapted by evolution.[35]

Additionally, exposure to nature has been studied as a variable affecting cognition and mood and as an intervention in mental health disorders. Many studies document benefits on cognitive function and mood, both in urban green space and in wilderness settings, though the exact "dose" and what kind of nature exposure is needed for specific benefits remains unclear.[36] Salivary markers of stress decreased most markedly after 20–30 minutes in a natural environment, suggesting a relatively short exposure is beneficial for stress reduction.[37] Gardening and other nature-based therapies generally have shown efficacy in managing depression and anxiety, with improvement in symptoms and higher rates of return to work. Incorporating agricultural activities in the setting of "enforced residential populations" such as prisons and the military, enabling people to work in nature and see the results of their efforts, has been shown to improve their sense of purpose, cooperation, and tranquility. While objective measures constitute ideal scientific metrics for analysis, powerful testimonials about the effect of nature on individuals' outlook, mood, and sense of agency also align with the concept that nature offers life-changing rewards to people at the extremes of human stress.[38] This has been demonstrated by the success of horticulture programs in shelters for victims of domestic abuse and in farms incorporated into correctional facilities.[39] Prospective randomized trials of greening empty lots in urban neighborhoods by planting trees and grass, compared to just picking up trash or having no intervention, have been shown to reduce the incidence of neighborhood residents feeling depressed or worthless by nearly 50 percent, with even greater effects in low socioeconomic areas.[40] These results suggest a powerful, simple, and cost-effective application of nature that demonstrates major benefits.

Public Health

Nature's benefit to public health can be harder to study, because the researcher doesn't have a "captured population" offered by a cohort of patients with a specific illness or in a specific living situation. Additionally, "exposure to nature" in day-to-day life is more difficult to define and measure than whether your hospital room is one type or another, and measures of health cover a lot of ground in many different domains. Nonetheless, information may be collected using surveys, demographic data, and longitudinal studies that follow subjects over time. Perhaps not surprisingly, a significant body of evidence has accumulated linking exposure to nature with overall physical and psychological health in people of all ages and across a wide diversity of locations and cultures.

Researchers studying children's health in Spain looked at associations between living close to parks or forests and the incidence of obesity or being overweight, time spent each day on computers or other screen devices, and asthma or allergic symptoms. The idea was to look more objectively at potential health benefits as well as potential risks—not to be overlooked—of living in close proximity to nature. On average, children living closer to outdoor green space as determined by satellite maps spent less screen time and had a lower incidence of being overweight. Those in proximity to forests had no greater incidence of asthma or allergies, while those living next to parks had a slightly increased incidence of allergies.[41] However, this kind of population-based study does not measure how much time kids actually spend outdoors. And because the beneficial associations with health metrics and green space proximity are greater in children whose parents are more highly educated, parental influence may be playing a role in these outcomes. Nonetheless, this type of data has spawned a trend for health-care providers to advocate for and even to "prescribe" outdoor activity for children, in part to counteract the increasing time spent looking at electronic media as well as the reduced access to wooded and open spaces for unstructured play in many urban and developed areas.[42] We will encounter more on this subject in Chapter 9, when we discuss the Green Children's Hospital project.

An impressive array of studies across a range of behaviors and benefits show similar associations. For instance, children from low-income urban environments who have a yard or other nearby green space to play in on average have higher cognitive function than children in high-rise buildings with no outdoor play.[43] Adolescents rate the opportunity to be outside in safe green space as highly important to their sense of living a healthy life, and those who participate in outdoor physical activities have higher health-related quality of life in physical and emotional functioning when followed over time.[44] Participatory involvement with the outdoors and affiliation with animals correlate with well-being in adolescent girls in Finland.[45] College students make less impulsive choices when exposed to natural—as opposed to built or abstract—visual stimuli.[46] Children in Denmark consistently exposed to residential green space below age 10 have a lower incidence of mental health problems in adolescence and adulthood, even controlling for other risk factors; this is especially true for urban residents.[47] Walking for 30 minutes improves health measures, but walking in a natural environment also shows additional cognitive benefits compared to walking in a built urban environment.[48] In a study of 10,000 Dutch residents, close proximity to green space, such as a household garden, was linked to improved self-reported physical and psychological health metrics, especially for those in lower socioeconomic groups and those spending more time at home.[49] We even live longer with nature. Elderly adults who live near green space participate in more physical activity, and those in urban areas who live near parks and vegetation have increased longevity, possibly because of increased walking and socializing.[50]

There are exceptions to these positive results. No relationship between proximity to green space and emotional well-being was found in young people in Canada, except for a weak association for those living in small cities, and none was found in Singapore.[51] A study in Norway on what factors were associated with adults participating in physical activity in natural environments found that childhood experiences in nature, and the ability to undertake activities with other people as part of a social network, were the strongest predictors—suggesting that nurture and the opportunity to socialize, rather than an inherited love of nature, reflect important factors in people's behavior.[52] And in some areas of South Africa and Australia, living near parks *increases* stress—because parks are perceived as unsafe.[53]

Despite these potentially compelling arguments, the idea of biophilia as a reflection of an inborn genetic predisposition has met with considerable skepticism. Some authors criticize the concept as well as the evidence as sloppy, lacking the objectivity and controls essential for conclusive science.[54] Others cite counterexamples. Some Indigenous persons prefer jungle terrain to a savannah landscape. Some cultures don't seem to universally fear snakes, they just fear the poisonous or dangerous ones; chimpanzees, our closest primate relative, are more afraid of turtles than snakes.[55] The preference for open, park-like landscapes isn't universal and decreases with age, when people equally prefer landscapes that are familiar.[56] What about the fact that people tend to migrate to cities, rather than the other way around? Nonetheless, it goes without saying that neither examples nor counterexamples provide "proof" of anything; they are supportive or contradictory observations on the complex network of factors that influence human behavior.

Other opponents view social learning and culture, rather than genes, as most important in shaping people's attitudes toward nature and believe that much of this relies on expediency.[57] People engage with the natural world just to the extent to which it is necessary to meet basic needs. Nature is more frightening than soothing in many places and likely was threatening through most of human history.[58] Depending on the context, say these scholars, people know and respect and are affiliated with nature to the extent to which that knowledge serves their overall short-range purposes. If this were not the case, the tragedy of the commons would not occur, as people would respect and love nature enough to protect it.

Some criticisms of the biophilia hypothesis appears to arise in part from a distaste for the notion that genes "control" behavior rather than that humans make choices based on perceived advantages within their immediate circumstances.[59] Here also the details provided by neurobiology can provide a more nuanced perspective. As we've seen, the brain isn't "either / or"—it isn't "controlled" by genes but can be predisposed by genes. It isn't that something is learned or genetic—it's virtually always both. It isn't that culture or context is fixed and we "are" one way or another—we are designed to change and adapt, but not too fast, and not at the exact same pace as others in our society—there's a distribution, *by design*. It's the interaction of predisposition and learning and experience and our current circumstances in those 86 billion neurons, each with their 10,000-plus synaptic connections, that causes us to be swayed. Countless present and distant, logical and emotional, shifting minuscule influences rank ordered by importance at that instant result in a given behavior at a given moment. Yes, if biophilia were a strong and primal urge in everyone above all else and under all circumstances, we'd be in a different place. As it stands, the evidence seems to suggest with some consistency that most of us do have an affinity for nature, that exposure to nature seems to have a salutary effect on many aspects of learning and emotional well-being, but also that it competes with many other priorities in our individual and collective decision making at any given moment.

Eco-depression

If the time frame for the threats embodied by the Anthropocene were eons instead of decades, the genetic blueprint directing our neural equipment might have been selected to perceive and react with the appropriately urgent response to what is now a genuine crisis. But the history of humans and the planet has been characterized by a perfect storm of coincidences that each contribute to the situation in which we find ourselves. First, human beings' superior cognitive and problem-solving abilities have made us the most successful "weed species" of all time, colonizing the entire globe and reproducing at a dizzying and accelerating pace, in a very short geological time. This success, when it is actualized in the context of relative affluence, is something that we are predisposed to feel as *positive*—we are fed, and healthy! We have found new suitable places to live! New babies in the family—how wonderful! When most of the contemporary world's rapid population increase happens outside the immediate experience of those in the high-income regions, it is more difficult to perceive it as threatening or negative. If you live in a suburb in the United States, you don't experience directly what is happening, population-wise, in Bombay or Lagos.

Land use for all those people comprises another major environmental strain that may not be perceived by those in the high-income industrialized world who produce the most carbon emissions per capita. Our increasing population develops more and more land, squeezing out more and more biodiversity. But most of us don't see this directly in ways that have visible, immediate consequences that we are designed to perceive as threatening. Extinctions happen slowly, often to species with which we are unfamiliar. We still have squirrels, and robins, and enough bumblebees that most of us who spend most of our time in office buildings or inside our houses don't see changes in ways that alarm us viscerally.

And carbon dioxide? We don't have the bodily equipment to perceive that this now has doubled over the average level present during all prior human history, and has gone up by another third just during the lifetimes of those of us born in the mid-twentieth century.

So, according to what our brains evolved to warn us about and reward us for doing in order to survive, on the whole we might expect to be pretty content. And, in fact, going about our daily lives, with respect to what tends to clamor for immediate attention, environmental concerns end up being pretty low on the list. When viewed from a brain design point of view, this should not come as a big surprise.

But still. Even here in the United States, what is that vague visceral dread some of us feel when we see another patch of forest turned into chain-sawed trees and tractor-carved mud for yet another housing development? When walking along the shore means picking your way among half-buried plastic bags and a garish rainbow of human-made flotsam and dead sea life and chunks of Styrofoam? When we face the monarch smashed on the car windshield, the oily sheen of a lifeless stream, the 72-degree day in February in New England?

And it's not just adults with a particular point of view who feel this way. Kids have it too. The second grader in my urban neighborhood whose mom told me she cried for an entire day when the six evergreen trees on a lot near her house were cut down. Or the girl in Richard Louv's book *Last Child in the Woods:*

> "I had a place. There was a big waterfall and a creek on one side of it. . . . I'd look up at the trees and sky. . . . I just felt free; it was like my place . . . I used to go down there almost every day."
>
> The young poet's face flushed. Her voice thickened.
>
> "And then they just cut the woods down. It was like they cut down part of me."[60]

This sentiment has been called "eco-depression," "climate change depression," "solastalgia," or even "The Anthropocene Blues."[61] That concern about climate change and ecological decline is an increasing and worldwide phenomenon has been documented in numerous studies, from Papua New Guinea to Australia to Switzerland to Canada. Obviously people whose way of life and livelihood are directly affected or whose lives are endangered by forced migration from direct disasters like sea level rise or droughts or wildfires have the greatest immediate claim.[62] A form of "climate post-traumatic stress" has been described for those whose lives have been upended by these direct catastrophes.[63] Those of us who witness this but are not directly affected may feel deep empathy for those who are suffering. But also young people around the world who feel that their futures are uncertain and who have little political power also experience distress, because of the "post-ecological dilemma" characterized by the presumably insoluble conflict between environmental goals and consumerism.[64] A sense of helplessness has led some authors to posit that we may respond to this dilemma by just giving up, and that some are already in Kubler-Ross's advanced stages of grieving.[65] Worries about the future have been termed a form of "pre-traumatic stress disorder," a vague threat of a looming crisis.[66] Others have pointed out that habitual worrying about climate change is not a mental health issue but is grounded in reality and, if anything, is a rational response—one that may spur effective action.[67] The worldwide youth Sunrise Movement is one such social epidemic. Perhaps more than biophobia, eco-depression may serve as a corroborating testament to the existence of biophilia, an indicator that threats to the natural world cause distress in humans, with components both of cognitive and emotional impact.

Biophilia and Climate Change

For all but the most recent epoch of human history—the very end of the last step on our timeline—we were surrounded by and integrated into nature virtually all the time. The natural world *was* our world, with some accommodations; the idea of living largely "apart" from nature would have seemed preposterous to earlier humans. Since our brains evolved at a time when continuous exposure to the natural world was the norm, we simply did not need to evolve a five-alarm warning signal when we were separated from nature, like a fish might thrash when out of water—get back in, RIGHT NOW, or you'll *die!* No, we can function in the modern world without much personal exposure to nature at all, since our societal systems and adaptations allow our basic needs to be met just fine without it—for the short term. And being chronically separated from the natural world happened in modern life in too short an evolutionary time scale to evolve a warning system, just like we don't have a built-in red light for carbon monoxide poisoning, or toxic chemicals in plastics, or climate change. Nature *is* rewarding—that's why people vacation in it—and it can be a powerful reward, transformative even—but it's a quiet, subtle reward, not the immediate, survival-driven, instant dopamine surge of warmth when you're cold, or food when you're hungry, or an attractive mate under the right hormonal conditions, or personal agency and accomplishment. No, for most people, nature feels more like a choice rather than a need, a gift to oneself, a neural balm rather than a *zap*.

But there is also considerable evidence that exposure to and affiliation with nature can serve as a catalyst to environmentally sensitive behaviors. Both love of nature and eco-depression have been the critical ingredient that has changed some people's priorities and decisions. This is more true for some individuals than others, for reasons that remain incompletely understood. In some people, biophilia is a core driving force for much of their decision making. For some brains, this element is the behavior-change agent that can tip the scale. The science shows that those who feel this strongly can't assume everyone else does—there is considerable variability.

If we love nature, why can't we just muster the will to save it? In part, it's because it's only one part of the complex web of inputs we use to make our second-to-second decisions. The variability in this trait suggests that for some people, the strength of this element among the millions of considerations that influence their decisions will prevail. For others, the relative strength of other considerations will override their love of nature. Think back to our legislator. A personal love of nature may be one component that plays a role, but how the decision ultimately concludes is based on a tsunami of competing weights, some logical, some irrelevant, many subconscious.

Recognition of biophilia and its inverse, eco-depression, may be one tool affecting the equation that gets some people to move from reacting viscerally to nature or its destruction, to taking action on its behalf. The drive to protect something you love, especially when it is threatened, can be a powerful force. Research of what works to "cure" eco-depression is sparse, although the principles of agency would suggest that taking concrete action on behalf of something valuable that is at risk may help.

So can this knowledge help in the campaign to change how we live, individually and in our collective priorities? We will learn more about what researchers have discovered about affinity for nature and pro-environmental behavior when we journey through the landscape of changing behavior in the next chapters.

But the other part of the answer to why we can't just use love of nature to save it arises from the fact that often *we don't know what to do.* Without knowing which behaviors to change, and how they interrelate to consequences that might directly affect those aspects of "nature" we care most about, the connection between our behavior and the outcome remains tenuous—a setup for reward deficit. It's challenging enough to save a wooded lot from being bulldozed in your neighborhood, never mind saving the planet.

To understand better how all these factors interrelate, we need to understand how modern life interacts with our brains to influence contemporary choices and behavior in the twenty-first century. And which, specifically, of our behaviors needs most to change from an environmental impact perspective, both individually and collectively? These topics are the focus of the next chapters on our journey. We will then be able to see how biophilia may fit into the overall environmental equation from a neural point of view.

Part 2

The Twenty-First-Century Brain

5

An Acceleration of Consumption

A DAD PUSHES A BABY IN A STROLLER. He is talking away as they move together down the street at a brisk clip. But as he moves past, you realize he isn't talking to the baby. He has an earphone and is conversing animatedly with a client at work.

A technology company executive travels across the ocean to attend three meetings in three countries in as many days. The meetings facilitate progress toward commercializing some new software to improve data security. She is just starting to adjust to the change in time zones when she returns to a full schedule at her company headquarters on the sixteenth floor of a steel and glass office building in a large city.

A nine-year-old works in school on a tablet computer. After school, she and her friends communicate from home via a computer game in which they adopt and care for virtual animals. She has strong preferences about what clothes she wears, what brands of food and toiletries the family purchases, and even what car is used to drop her off at school, based primarily on inputs from her peer group, television, and social media, and she feels keen distress when she perceives her brands aren't the favored ones in her social group.

* * *

So far in our journey, we have looked at the evolution of the brain going back, through millions of years, to simple organisms. We next explored the functions and types of rewarding experiences and how we learn from and are motivated by reward. The examples above illustrate another factor we need to take into account as we examine neurobiological influences on our environmental crisis—how our brain design may interact with our daily lives in wealthy industrialized societies to further promote the *acceleration of consumption*. Because our brains are continuously modified by what they encounter in our individual experience, modern life in high-income countries clearly has changed our brains. The adaptation of our inherited reward system to changes in modern life may promote behavior with increasingly negative environmental impact. To explore this further, let's take a look at the incredible adaptability of the brain.

From Hominins to Today

The story of human brain evolution historically relied on several lines of evidence: measuring changes in brain size based on prehistoric skulls, inferring behavioral capacities from tools, and extrapolating from the conditions in which early hominins (prehumans) lived what the cognitive capacities their brains would have needed to survive.[1] But the clever ways in which more recent scientific techniques including molecular genetics and advanced neuroimaging have elucidated some specific genetic changes influencing the size, structure, and complexity of the brain have shed new light on how—and sometimes why—our brains have adapted and changed over the history of our species.

To understand this, one needs to understand a basic and fantastical characteristic of the brain—*plasticity*. This characteristic is fundamental to the human success story. What this means that the brain—especially the human brain—is able to expand the area that it dedicates to a specific function on demand. For example, people who are born without sight because of a problem with the eyes often have extraordinary hearing. This occurs because the brain volume that would have been used for vision now gets allocated to the other senses—which, having more brain assigned, may be more acute than in people with normal vision. In another example of how plasticity occurs, monkeys who undergo surgical removal of the area of the brain that controls the movement of one of their fingers will gradually "condense" the parts of the brain that control the adjacent fingers, reallocating some of their own brain area to the finger associated with the brain area that was removed. This plasticity allows the animal to recover nearly normal use of the hand—the same way that humans recover from a stroke or an injury. The ability to heal and also to adapt to changing circumstances are amazing features of living organisms. The human brain exhibits these traits in ways that we often take for granted but are essentially miracles of adaptability that we experience on a daily basis without being aware of them.

Apart from illness or injury, people in their everyday lives change their brains with practice. London cabdrivers have larger areas of their brains allocated to spatial skills that correlate with the number of years of experience.[2] Conversely, relying on global positioning system devices to get around causes your ability to use a map and figure out where to go by real-life visual cues to diminish and the navigation area of your brain to shrink.[3] Highly skilled musicians have larger areas of their brains allocated to the tasks needed for their performance, as do athletes, as do people who text with their thumbs—those brain signals get stronger on a day-by-day basis, depending on how much one texts.[4] Learning a new skill and doing it a lot changes your brain physically—that's plasticity. Likewise, the old saying "use it or lose it" also is true; when you stop doing something, you become less adept at it. On a cellular and molecular level, this occurs by the mechanisms we encountered in Chapter 2—strengthening of new connections, reinforcement of newly linked-together networks, and allocation of areas of brain to the frequently performed activity by enhancing connections to other neurons engaged in that task.

But why not just stay at tip-top form for anything you've ever learned, forever? *It's because you only get so much brain.* It has to fit in your head and efficiently use the calories and other nutrients you supply to do its work. So, when you're not using connections, their constituent parts reconnect in new ways for those things you are learning and doing in the present, to serve you best right now. In fact, besides making new connections, your brain also constantly regenerates in specific ways, by creating new brain cells and replacing old ones, in a few specialized places. In the hippocampus, the memory hub we met earlier, the formation of new neurons is specifically correlated with *forgetting* (at least in mice, where this has been tested).[5] Forgetting is an important task, to help you learn that new password you had to come up with for your computer account. Imagine if you couldn't forget anything! This is the circumstance for a few rare individuals with what is called "superior episodic memory"—and those folks don't necessarily consider this a terrific thing.[6] Instead, remembering what's useful and forgetting what's outdated works best overall. For most of us, plasticity, including the ability to forget what we no longer need to keep in mind, is how we manage to get through an entire lifetime with the same brain, that's efficient enough to fit inside your head.[7] That the *new* activity or information you learn must be rewarding in some way is a given—it must help accomplish tasks for basic survival or otherwise provide a benefit in daily life, or you wouldn't be able to learn it at all. Reward is the way your brain figures out what to learn to help you live and reproduce—that's what it's for.

We developed the capacities we have as humans because genetic changes that happened at random gave us an advantage and thus persisted in the population. Human brains are large in specific places—in particular, in the frontal lobe, where much occurs that gives us some of our survival edge, such as analytic and decision-making skills and components of language. Like the reward system clues from Parkinson's disease we learned about in Chapter 2, scientists have been able to piece together the kinds of changes that facilitated the evolution of language based on genetic mutations that cause language deficits in specific families and communication problems in birds and mammals.[8]

But as brains enlarged and capacities increased, our human-like hominin ancestors gained increasing capacity to *alter their own environments.* Scientists speculate that human-caused changes in how people lived in turn led to evolutionary selection for specific brain adaptations that could best function within these new societal paradigms. For instance, in a large brain whose size allows for changing functions of the cortex, depending on the experience of the individual, behaviors like tool usage would correlate with expansion of the brain circuits engaged in that activity, in turn facilitating increasingly sophisticated tool use and language.[9]

According to this theory, the enlarging brains of our ancestors—along with the inherent plasticity that allows adjacent brain regions to expand to take on new tasks—gave humans the exact equipment that could best adapt to changing evolutionary pressures created by humans themselves. For example, the formation of the ability to pass on information and skills to others efficiently by utilizing language instead of gestures likely led to advantages for specific groups who embodied these capacities.[10] In this way, small genetic changes, coupled with plasticity, reward, and natural selection, allowed for the gradual evolution of new functions in humans that matched the needs of the environment we lived in and that was altered by our own human interventions.

The Acceleration of Change

The behavioral evolution of tool use and language took millions and hundreds of thousands of years, respectively. Cultural and societal changes in prehistoric time also occurred over millennia. For instance, the current array of major religions generally are thousands of years old. While the forms have changed, institutions considered basic to most societies—agriculture, urban and rural living, universities, libraries, marriage, child-rearing—all, it can be argued, would be recognizable if we could time-travel back to many points of early recorded history.

But science and its offspring, technology, build synergistically on intersecting advances. For this reason, the pace of change in technical fields continues to accelerate at an ever-increasing rate, which in turn leads to cultural changes. Since science and technology now change much more quickly than biologic evolution, cultural changes are happening also increasingly quickly, whether our brains are adapted to that fast a pace of change or not. For each generation, the differences between what life was like when a person was born compared to what it is like in middle and old age are larger and larger. As one example, language dialects and terminology used to stay relatively constant over many generations, evolving steadily but gradually. At present, if you can keep up with your kids' slang words for more than a few months, you're doing better than average.

As we learned in Chapter 3, humans are rewarded by novelty, but we also crave familiarity. Some authors have hypothesized that the pace of change in the postindustrial world is so fast that people's brains struggle to adapt. In his famous 1970 book *Future Shock,* writer and futurist Alvin Toffler, in collaboration with Heidi Toffler, suggested that people are increasingly burdened with malaise and uneasiness because of change occurring at too fast a rate to process and assimilate comfortably.[11] Change has only accelerated since the Tofflers' observation.[12] From a neuroscience point of view, as noted previously, in times of stress, the balance between novelty and familiarity gets tipped toward a preference for the latter for many people.

To see how the brain adapts through plasticity to the increasing pace of change and how these changes might have implications for the acceleration of consumption, let's look at what life looks like to a hunter-gather, and compare that daily existence to our contemporary lives in industrialized countries like the United States.

You live in a tent, a cave, or a hut. The day starts around the time the sun comes up. Your family sleeps together in one common space. You awaken to the sound of voices, birds, animals, wind, and water. The things you see, hear, and smell are your living space, and the outdoors.

Your activities are the same as those of generations before you. You wear what everyone else wears, that you or your family fashion from available materials. The expectations of what you are to do, of what you need to learn, and your daily tasks are determined by how things have been done for generations, as far back as people can remember. Information is obtained by observation and by the words and behavior of others, particularly older family members, leaders in the group and occasionally those coming from other places. There are customs for birth, illness, death, mate finding, child-rearing, punishment, praise, feast and famine, interacting with others, relocating, and celebrating. There are stories and songs. There are work tasks often done communally, in which children are included. You remain in your family group most of the time, throughout your life. You take on adult work gradually during childhood and become an adult as soon as you are physically able to perform adult tasks. You spend most of your time outside, retreating to your shelter only to sleep when it's dark or to ride out weather or other threats. You learn, through your brain's mechanisms for teaching you lessons specific to your circumstances, how to read the landscape, and other people. You learn and adapt and get better at navigating the tasks and challenges you face. Mortality is high, and people die and are born regularly.[13] Though some people live long lives, loss of family members in childhood and youth is routine. Illnesses and injuries often are accompanied by unrelenting pain and misery. Young mothers in labor whose babies get stuck in the birth canal die in agony by bleeding to death. Infectious diseases wipe out young children. Cuts or scratches that get infected, appendicitis, and tonsillitis are often fatal. Hunger is common; much time and energy are spent combating it. Work is physically arduous; nobody is obese. A successful hunt or harvest begets a party to which everyone you know comes. Hunger and survival are what drive most human mental and physical activity.

In the present-day average life in the high-income, industrialized world, likely familiar to many readers in this audience, you live in a house or an apartment. The day starts when your phone or alarm clock goes off, before, during, or long after dawn. Your family members sleep in different rooms. You hear the whoosh or clink of your heating system, your television, radio, the electronic jingle of the microwave, the refrigerator humming, traffic noise, sirens, airplanes. You smell indoor smells, perfumed products, plastic, and vinyl. You may spend the entire day without going outside at all.

The day may start with electronic media showing rapid-fire images and sounds of news and opinion, entertainment, and tragedy from around the world. Food, often in forms unrecognizable as arising from nature, is obtained from the cupboard and refrigerator, water from the sink, requiring little if any community effort. People have separate bathrooms.

The family segregates. People commuting have earphones and a continuous soundtrack. Kids watch videos in the car on the way to school, then separate by age. Formal schooling is mandatory and lasts for many years. Parents go to work, far from other family members. Most children do not observe parents' work in a way that teaches them how to perform it. Grandparents, aunts, uncles, and cousins live somewhere else. Community leaders are not known personally. Major illnesses and mortality in the young are rare. Anesthesia makes most procedures painless. Medication cures most infections. Death is generally managed with interventions that mitigate agonizing physical misery. Most people have ways of staying safe from the elements. The majority of people has enough food and water to sustain life.

Light is everywhere, day and night. Information comes electronically, constantly, from around the world. Exposure to entertainment and advertisement, which is highly dense with respect to sensory input, takes up hours of most days and is challenging to avoid. Science and technology advance day by day. Most people are sedentary and remain indoors for most of the time; on average, 90 percent of your time is spent inside a building.[14] Behavioral norms change rapidly and vary among people. Tribal affiliation is more likely to coalesce around shared entertainment preferences and sports teams than interdependence for group survival. Beliefs about truth and falsehood, right and wrong, expected and acceptable behavior differ between grandparents and parents, parent and parent, parent and child, siblings, and friends. Religion is heterogeneous, and one set of rules contradicts another, with adherents coexisting in close proximity. There are few if any shared authorities.

In the world of work, reward comes in the form of a paycheck at regular intervals. It is related to the work performed, but cause and effect are indirect compared to hunter-gatherer rewards. There may be praise or recognition, but this too is typically after the fact. In many work contexts, skills become outdated rapidly. Staying afloat, getting ahead, dealing with problems, and planning future activities take up more mental room than imminent starvation or community survival.

These dramatic changes in lifestyle have occurred in an overlapping time frame with advances in science and technology as well as the acceleration of per capita consumption in the developed world. We consume more energy, information, and resources and have more material goods than ever before. To explore how these changes in technology and lifestyle might be related, let's next look at a few areas of modern life that have been studied in more detail, and how these may intersect with brain plasticity as well as behavior relevant to consumption and the environment.

How Children Spend Their Time

Researchers have tracked significant shifts in how children spend their time over recent decades, a millisecond of evolutionary time. These changes on our forty-day transcontinental walk timeline represent only a fraction of the Anthropocene duration, reflecting the extraordinarily rapid acceleration of change. Both the average human lifespan and the duration of what is considered "childhood"—especially adolescence as a subcategory—have become greatly expanded since hunter-gatherer times. These general changes in societal concepts of childhood evolved in the postindustrial age and accelerated during the second half of the twentieth century. Recall the theories of learning and evolution described in Chapter 4, which suggest that unstructured play is how children's brains have been designed to learn skills, cooperation, and interpersonal exchange. Surveys measuring changes between 1981 and 1997 in the United States showed that children spent less time in free play and more in structured activities as those decades progressed.[15] American children in 1997 watched on average over an hour and a half of television per weekday and over two hours per weekend day; only 25 percent of children played video games in 1997, but of those who did, the average amount of time spent was just under an hour per day.[16] By 2005, a national survey on electronic media use by preschoolers in the United States showed that on average, preschool children spent an hour and twenty minutes watching television or video media, and five- and six-year-olds spent fifty minutes on a computer.[17] A 2009 survey of mothers from sixteen countries around the world confirmed a decrease in free play compared to what the mothers recalled experiencing during their own childhoods, and an increase in television viewing, which had become the major free time activity.[18] By 2012, data revealed that children spent over two hours per day on television and over an hour per day on other electronic media; preschool children clocked in at four hours of television per day.[19] Child television and screen time correlates with parent time spent in similar activities, increasing also with decreasing socioeconomic status and urban location, and with having a television or computer in the child's bedroom.[20]

The exploding use of multifunctional mobile phones for communication, information, and entertainment has further complicated the assessment of "screen time," which appears to have increased markedly as cell phones have become ubiquitous for younger and younger children. Screen time via mobile electronic devices is now interspersed temporally and spatially with all other daily activities; it no longer is a separable leisure activity restricted to a specific time or place. Researchers have studied the effects of screen time on child development, sleep, adult behavior, addiction, and other behavioral changes, including both social connectedness and social isolation; the effects are complex, occasionally positive, but often deleterious.[21] And it's not just the kids. Parents who use smartphones when they are spending time with their children in playgrounds, museums, or restaurants have been observed by researchers to pay less attention to and develop less connectedness with the child.[22]

The phenomenon of high-frequency use of cell phones with features of behavioral compulsion, or even reaching criteria for addiction, has been described using a name appropriated from the 1980s, when widespread use of computers in the workplace became commonplace: *technostress*.[23] As initially defined, technostress reflected people's difficulty accepting and successfully mastering new technologies in their jobs. Several decades later, the term has morphed to include compulsive overuse of technology and anxiety about being "out of touch"—a remarkable progression from resistance to new technologies to acceptance and a widespread perception of the *necessity* of minute-to-minute use of technology in everyday life. The fear of missing something has been described in adults, but may be particularly acute in children and adolescents preoccupied with peer acceptance.

Because these changes are happening so quickly, organizations such as the American Academy of Pediatrics have had to scramble to try to keep up and to provide frequently updated recommendations for parents.[24] While the presence of televisions in children's bedrooms has declined, the effects of constant access to social media inherent in smartphone technology, and the increasing number and decreasing age of children having phones, are influencing many aspects of health and wellbeing that researchers are racing to understand.[25] These changes in technology access also are occurring in the context of other social changes, including working parents, nontraditional family arrangements, the so-called "24-hour news cycle," and the use of digital media in education. This makes it challenging to separate entertainment / play from school assignments / work. Compared to when it's just time to put the Legos away and do your homework, it is difficult to limit one use of technology without penalizing another. The "But mom, I need it for schoolwork" excuse adds a different flavor of technostress to parents trying to do the right thing for their children, with virtually nothing to guide them from accumulated cultural experience—it's just too new an issue.

Recent years have uncovered even more uncharted territory in the digital debates. The collection and curation of information—and deliberate dissemination of misinformation—via wide-reaching social media, the determination of what information you see based on a "profile" created from your browser history or personal data that may have been bought and sold without your knowledge, and the anonymity of many social media platforms all influence the information received by individual people that modulates our perspectives, associations, and beliefs.[26] These interactions can barely begin to be studied before they change into something else new and unforeseen. Change is happening month by month, not year by year or generation by generation or century by century. This pace of change itself is interacting with our brain equipment to propel—and perhaps unsettle—our reactions, decision making, and behavior. Of course, it may also expand our capacities and lead to valuable social and scientific advancement. One effect does not preclude another, but advantages may come at a cost that is impossible to predict in the absence of precedent.

How Electronic Media, Attention, Reward, and Consumption Are Linked

Reward circuitry is intimately connected to your attention networks and memory circuitry. The various platforms of electronic media are used in modern life both to teach skills and to influence consumer behavior. While the effects also are relevant to adults, we will focus on its effects on children, which have been studied extensively.

During evolution in the natural world, specific sensory stimuli had survival value. Things that moved, looked lifelike, and made sound might mean prey—or a predator. Things with bright colors might mean food. Thus, during natural selection, our visual system was gradually refined in its design to be biased so that motion in our peripheral visual field, for instance, is in effect "magnified" in our brains such that it immediately and effortlessly draws our attention. Think of how you notice a fly in the room without consciously looking for it. Likewise, things with features that look lifelike, sudden distinctive sounds, bright colors—each of these characteristics has been shown to catch the attention of both children and adults. This doesn't happen by accident; instead, it's a function of the way the eye, brain, and attentional neuronal networks are connected and "weighted" such that these survival-value stimuli like sudden motion are "exaggerated" in your perceptual circuitry, inherited from the evolution of the nervous system for the particular circumstances of our species. We take our perception for granted, but it's not camera-like—it's filtered through our nervous systems to make things we need to pay attention to even more noticeable and interesting to us. These specific design elements of how our nervous systems collect, process, attend to, enhance, and act on sensory input that is relevant for survival have been worked out in great detail in a variety of species, from different types of flies with specific hunting strategies to rodents to monkeys to humans.[27]

Furthermore, with respect to the content of sensory input, researchers have shown that if the stimulus provides clear, comprehensible, and relevant new information about the world, this further enhances its reward value.[28]

The first widespread use of electronic media to enhance learning in children was "Sesame Street," a rapid-fire format for early childhood education. This public television show used characters, animated letters and numbers, and sounds, all packaged in multisensory stimulation—light, motion, sound, space—that was designed to capture and hold the attention of preschool-aged children.[29] Thus, the creators of educational programming brought all of these evolutionarily determined attention-grabbing features together in one vehicle, with a sensory density of sound, motion, color, action, and contextual relevance, all concentrated in time and space more than anything that occurs in real life. Think of a cat going after a laser pointer—the intense motion and brightness outcompetes the most sparkly and energetic bug or shiny-feathered bird, capitalizing on kitty's instincts to respond to those features, and compelling her to attempt relentlessly to capture the uncatchable.

Early studies showed that the Sesame Street approach was a big success for its intended result. While young children's attention span for conventional preschool activities was typically measured in minutes, kids watching Sesame Street remained in eye contact with the television screen for more than 80 percent of a one-hour program.[30] And when a person pays attention to something and is rewarded by it, learning occurs. Since television is so widespread, the hope was that this attractive programming would give a boost in school skills even to children living with socioeconomic disadvantage. A significant body of research utilized different versions of Sesame Street that manipulated specific features in order to understand what elements of sensory stimulation best captured and held the attention of children of different ages.[31] Such data could be used to inform well-intentioned educational programming. Parenthetically, it also could be used by companies seeking to influence children via attention-grabbing advertising, influencing consumption behavior for a young target audience, as further discussed below.

Data also began to accumulate that exposure to television during childhood decreases children's ability to pay attention in other life contexts, as judged by their parents.[32] Thus, some educators argue, how can we expect children's brains to pay attention to stationary black letters on a white page, as occurs in standard reading in a classroom, when they're used to being entertained during learning by dancing letters, furry singing characters, and short dense bursts of technicolor informational entertainment? How can reading a book compete with video games, or product placement during animated television shows? Will the cat go after the bug if distracted by the laser pointer? Or will cats exposed to laser pointers get even smarter, and catch bugs even better? And maybe bugs and birds are irrelevant to most modern cats anyway. While no one argues that learning to read is irrelevant, newer ways to educate children using visual images might be more effective and relevant to modern life. Because this whole topic is new and has no past to rely on for prediction, it has opened a whole universe of debate, and topics for research, all in a social and technological milieu that is changing by the minute.

There is considerable evidence that early life exposure to electronic media, with sensory stimulation features similar to those studied extensively in Sesame Street, creates measurable changes in children's behavior that can be adaptive or maladaptive, depending on the context. The specifics have changed as media use has morphed from traditional broadcasts on television to newer smartphone and computer-based digital information. For instance, in 1970, children began to watch television regularly on average at about four years of age; now they begin using electronic media at four *months* of age.[33] Even infants in the first months of life readily learn to use touchscreens.[34] Kids can learn preschool skills that carry over to the school years from electronic sources, but before age two, they still learn better from interacting with caregivers.[35] For this reason, the American Academy of Pediatrics has recommended that children younger than two years not engage in electronic screen time at all.[36] Older children can learn some aspects of problem solving and strategy from electronic media, and some researchers and commentators suggest these skills are well adapted to the digital age in the work world.[37] Parents' own use of and attitudes toward technology correlate with the extent to which electronic media are used by young children, and how much parents perceive this as potentially educational and beneficial.[38] The term "gamification" has been used to reflect how much of learning as well as entertainment has been turned into games that are ubiquitously and continuously available.

But other studies have shown that children exposed to electronic media can have more difficulty paying attention to other things, can exhibit more aggression and behavioral problems, and report more somatic symptoms like headaches and disturbed sleep.[39] The exact causes of these associations are incompletely understood; while investigators have tried to control for many potential confounders, such as preexisting influences, it hasn't been proven definitively that the exposure to electronic media necessarily causes the differences. But a large body of data remains compelling. In a study of six-month-old infants from lower socioeconomic backgrounds, those who had greater exposure to electronic media demonstrated poorer cognitive and language outcomes at age fourteen months, regardless of its educational content.[40] A European study of over 3,000 children found increased television and video time during the preschool years correlated with an increased risk for social and emotional problems when children were followed over two years.[41] A study of 400 middle school children showed better visual-spatial skills but lower grades for children with high screen time.[42] Other studies have shown alterations in the thickness of various cortical regions involved in decision making and attention, as well as changes in reward circuitry function, in adolescents with online gaming addiction.[43]

Despite this kind of evidence, it is difficult for families to scale back media use, even for young children. Recognizing this refractoriness, public health programs have been instituted to help families cut back on preschoolers' viewing time.[44] But what's on screens remains intensely interesting to children. In my own practice, and in those of other clinicians who treat kids, coaxing a child in your office to turn off the video game when it's time to be examined most often becomes not just a request but a battle. Children genuinely become dysphoric when the game is turned off and the device is sequestered by the parent, so that the physician can talk to the child, take a history about symptoms, and do a physical examination. This was less the case when children were brought to the office with a book or a stuffed animal. This difficulty doesn't mean that kids are inherently less nice today. Instead it is a testament to the carefully constructed attention-getting and reward features of commercial electronic media—the same features that make it challenging to do with less, and to avoid what some would label media addiction.

Does Content Matter?

In addition to the issues about how the format of electronic media affects the developing brain's information processing, attention, learning, and social development, concerns also have been raised about the content. How does what you view, in addition to how much you view, affect you later in life? A long-term research project about how viewing television in early childhood affected behavior during adolescence was published in 2001. Researchers showed that exposure to violent content during preschool predicted more aggressive behavior in adolescence. Likewise, educational programming in early childhood predicted better grades and more activities. While it's tempting to conclude that the content seen in preschool influenced the later behavior, the authors are careful to point out that it's also possible that children predisposed to aggression are more attracted to violent content, and children who are predisposed to doing well in school tend to like educational programming.[45] Some mild support was found by this same longitudinal study that the more time spent on viewing television, the fewer the number of activities and other creative endeavors undertaken by kids. Here too, however, the "chicken and egg" problem can't be fully overcome, even when correcting for factors such as parent education, because associations don't necessarily equal causation. Less active and creative children simply may prefer the low-effort engagement with a relatively passive medium. Finally, there has been considerable work on gender stereotyping based on programming content in television, games, and social media; in general, boys and girls respond to different content differently, which may both reflect and perpetuate stereotypes.[46]

Thus, content does seem to matter in shaping behavior and even beliefs. There is good evidence that exposure to video violence affects behavior, and that attention processing and distractability also depend on content.[47] That violent video games are rewarding has been confirmed by having adolescents play them while undergoing fMRI scans; researchers can see reward circuitry activated, most strongly when the subject is winning.[48] Electronic games are designed deliberately to be as rewarding as possible. As of 2016, it was estimated that three quarters of American adolescents owned smartphones, with 50 percent of them reporting feeling "addicted" to them.[49] Recognizing the difficulty in getting parents to cut down on children's' duration of electronic media usage, even at very young ages, some intervention programs have attempted to at least help parents modify the content of children's viewing toward less violence and more prosocial subject matter.[50] We will see how these changes may influence consumption a bit further in our discussion.

Life Cycle Changes, Age Segregation, and Adolescence

In hunter-gatherer societies, people of different ages were intermixed most of the time. But in modern times, another major cultural shift is *age segregation.* On one hand, older individuals in industrialized societies most often do not live in extended family units with adult children and grandchildren or other relatives, instead more often living alone or in communities of other unrelated older people. At the other end of the age spectrum in developed countries, children in general and adolescents in particular spend the majority of their waking time in groups of people of similar ages rather than in family or societal groups of mixed ages.[51] For children, the age segregation shift also reflects hours spent in school, where there are many children and few adults, and in social and recreational activities outside of school that also typically segregate by age. This means that learning by observation of the work and behavior of elders in their occupational and societal roles is supplanted by formal education in the classroom. Additionally, learning cultural norms and values occurs mostly among people in one's own age bracket. Sociologists and educators have pointed out that this gives outsized weight to peer approval, with major implications for social interactions, including purchasing and consumption choices. In adolescence, approval from one's peers, rather than approval from elders throughout a community, has become the most important yardstick by which adolescents and young adults judge themselves.[52] Fitting in, being accepted, and competing for optimal social standing in the peer group can be the source of intense pressure and not a lot of control.

In most of human history, adolescence as a concept did not exist. Children simply were born, grew if they survived, morphed into adult form, and took on adult roles. Relatively prolonged maturation during childhood helped facilitate passing on specific factual and cultural knowledge adapted to the widely different circumstances humans might find themselves in as they spread to different geographic and social niches. The remarkable plasticity of the human brain to change its skills, knowledge, pattern recognition, reward circuitry, memory, judgment, and attitudes based on experience and social interaction gave humans some distinct advantages compared to other apes. But the post-pubertal developmental period during most of human history was far shorter than it has become today.[53] Yes, there were adolescent initiation rites and ceremonies, but these were relatively brief, timed typically around puberty. The word "adolescence" wasn't used until the 1500s—an eye blink in the history of humans—and even then, the term was used to describe youthful behavior and preoccupations among individuals who already had joined the life of the community in an essentially adult capacity.

And the adolescent brain is, indeed, distinctive. It is now understood from advances in the science of brain development that emotional drives, typically concerned with seeking adventures, rank, and mates, often peak during the late teens and early twenties—the age of many of our oil entrepreneurs in Chapter 3. In contrast, the parts of the brain subserving judgment, impulse inhibition, and objective decision making don't complete their maturation (specifically, myelination, which speeds circuit communication) until people reach their late twenties to early forties.[54] This mismatch explains why adolescents and young adults often have a sense of invincibility and are prone to making risky choices that may end badly. While the evolutionary function of this discrepancy in the maturation of different brain networks is unknown, it has been hypothesized that it may exist so that young people are motivated by the lure of adventure and mate seeking to leave home and find new territory without being impeded by the hindrance of "knowing better." While adventure without sound judgment comes at the price of the loss of some individuals, on average, this is a necessary function, in humans and many other animals, for the dispersion and expansion of the species.[55]

In contemporary life in the high-income industrialized world, adolescence has become prolonged, often lasting one to two decades. During this time, individuals may undertake extended educational endeavors, live with some support from extended family, and work sporadically in serial occupational capacities. Most social interaction takes place among others of similar age, rather than within a mixture of ages. The age of first marriage has steadily increased; childbearing also occurs later than in previous epochs.[56] Individuals in this prolonged adolescence in high income countries often have considerable purchasing power, and peer influence can play a major role in consumer choices.

Add to this mix the advent of social networking. Here judgment of peers is instantaneous and large-scale—functioning as a sort of "mega-peer."[57] Social media can serve to normalize both prosocial but also antisocial or unhealthy behavior, and in this age-segregated social structure, the moderating influence of a community of elders generally is more remote.[58] Even things like speech patterns of media characters are imitated and adopted by children, apart from influences from real-world figures, such as parents.[59] Young children can so strongly identify with superhero characters that if they are wearing the costume, they don't always recognize that they can't fly, and sometimes jump off things to the point of suffering serious injury.[60] Over generations, who young people choose as "heroes" has changed from people they know in real life to people or characters they know electronically.[61] Among other factors, age segregation and the digital world have created an entire "youth culture" that didn't exist in the same way in the past and is more insulated from the wider community of people at different stages of life and different roles in society.[62] On the positive side, the digital world exposes people to other role models and points of view that have significant potential to broaden their outlook. Nonetheless, it has been noted to be increasingly challenging for parents to influence maladaptive media involvement by their offspring during a phase of brain development when self-regulation is immature, and when exposure is widespread, ubiquitous, and undertaken largely apart from direct parental oversight. Of interest, the most effective strategies for avoiding problematic electronic media dependency involve building strong interpersonal relationships between parent and child—in effect, bridging the wider divide that exists as a by-product of age segregation.[63]

To relate these changes to climate change, let's consider their effects on consumption. Among other effects, this synergistic confluence of changes in society and technology have contributed to the development of an enormous "youth market" commercial enterprise that exists to encourage consumption during this prolonged adolescent stage and to solidify robust consumption habits. Whether a product is healthy or harmful, necessary or frivolous—if it can be marketed in overt or subtle ways, by embedding it into images associated with being "cool" or "powerful," which is highly motivating for young consumers—it's fair game for this enterprise. Around the globe, enormous resources are invested in youth marketing because of the potentially high return.[64] Young children and adolescents (not to mention adults) are not always aware of the marketing messages embedded in entertainment and information and of their own co-opted role as brand ambassadors through their own postings on social media.[65] Another way to encourage consumption is to change what is portrayed as desirable at frequent intervals, which is fueled in part by the rapid cycling of news, cultural norms, and public figures acting as heroes to be emulated who are in ascendant popularity for shorter and shorter intervals of time.

It should be emphasized that there are many positive effects of digital culture. Researchers have found strong beneficial effects of positive use of social media among young people, from topics like cultural acceptance of diversity, citizenship and political activism, promotion of healthy behavior, and opportunities for service.[66] People of all generations can connect to like-minded individuals around the world, and many positive social movements—including those involving climate change activism, such as the Sunrise Movement—have relied heavily on social media connectivity; we will address this more in Chapter 8. Opportunities for education, access to carefully vetted factual information, and help with obtaining new skills all can be facilitated by technological advances in society.

Our focused exploration at present is not whether changes in twenty-first century living that rest on advances in science and technology are "good" or "bad" on balance. Rather, we are exploring the specific interaction between life in wealthy countries during this portion of the Anthropocene, and factors both brain-related and social that have contributed to the acceleration of consumption. As we learned in earlier chapters, it is the acceleration of consumption in its varied forms that fuels the acceleration of climate change.

Consumption Then and Now

Early humans consumed what was locally available, using every part of what they could produce or hunt, and later what could be traded. There was little waste, and considerable want. Throughout historical times, the wealthy few have used conspicuous consumption for "signaling" and political favor.[67] In contrast, in modern life in the affluent developed world, a staggering variety of goods is readily available, from all over the world, simply by looking online, and is available to more than just the wealthy few. Money exchange is virtual, by typing in a credit card number, or even simpler, confirming by a key click one that is stored electronically. Consumption is often private, so that neighbors and community members may not see your choices, nor weigh in as to their wisdom before the choice is finalized. A coworker likes cats, so friends and family can search out or impulse buy every kind of cat figurine, calendar, snow globe, pencil eraser, tee shirt, coffee mug, tote bag, mouse pad, beach towel, mittens, sticky notes, and slippers, appropriate for myriad occasions, appreciated but rarely needed in a survival sense. Waste also is segregated, so you don't see what becomes of your goods once they are discarded. Sophisticated advertising is everywhere, skillfully tapping into targeted and effective appeals for specific audiences, based on large bodies of data typically combed from electronic transactions.

Who has purchasing power also has changed over time. Historically (as well as in contemporary nonindustrial societies), children by and large were not the decision makers about what the family ate or how children were dressed. There just weren't many choices, and one ate and dressed according to custom and what was available. In contemporary high-income society, data suggest that children, influenced by media and their peers, help determine purchasing decisions not just for child-specific products like snacks, toys, and clothing, but also often for products used by the entire family like electronics, vehicles, and vacations.[68] The extent to which children influence purchases depends on cultural as well as socioeconomic factors, such as number of children, nuclear versus extended family structure, and parental employment.[69] For items like clothing purchased by adolescents, there has been a shift over past decades from parents providing the main influence on clothing choices to peers having the greatest influence.[70] This trend is consistent with the increasing influence of age segregation and the social media-driven "mega-peer" effect, such that children and adolescents are highly influenced by what they see on entertainment and social media.

In young children, concerns have been raised about media advertising and the ease with which manipulation can occur to influence what children want, and are prompted to ask for from their caregivers. In older children and adolescents, purchasing decisions can be influenced by brand sites on social media, placement of products in videos and movies, and perception of social cachet by association with celebrities and peers. Social media has been described as a vehicle for interpersonal persuasion on a mass scale.[71]

Rules made for print and network outlets that prohibit advertising health-risk items like tobacco and alcohol are murkier in the on-line world, where marketing may be overt or covert via brand placement and association with celebrities or other attractive people or settings.[72] Association of alcohol with risk-taking is another marketing nod to the adolescent brain.[73] Consistent with the function of the reward system to make associations, even when fast food companies voluntarily have scaled back on using toys as incentives and have included more images of healthier foods in their ads, children remember the toys and the less healthy but tasty foods that we evolved to find appealing when they were scarce and helped us survive—fat, salt, and sugar.[74] Indeed, because of the rapid change in technology and its intersection with society and norms of behavior, the term "creepy" has become used to denote conflicts between our historical sense of right and wrong and current applications of these concepts in entirely new and unchartered contexts of social media, advertising, and privacy.[75]

Why do people enjoy shopping? As discussed in Chapter 3, humans are rewarded by agency—the cause and effect of one's actions on an outcome. There are many things we can't control—often our jobs, our families, our society—but shopping is something immediate and personal. We find something, we make a choice, we get a result; we get something new. It's an instant gratification that can be intensely rewarding, at least for a brief time.[76] When done with others, shopping can provide memorable social bonding.[77] Other perks include competitive success, a sense of enhanced personal attractiveness, and group affiliation—evolutionarily weighted rewards, visualizable via tools used in the emerging field of "neuromarketing."[78] Debt—not tangible, hard to perceive, easy to discount—is a weak counterbalance.[79] Internet shopping reduces barriers further. That shopping is rewarding is illustrated by the fact that it can become truly, clinically addicting.[80] Like other addictions, an impulse buy often is followed by guilt, self-recrimination, and the need to feel better by doing it again.

But even short of high level, life-disrupting addiction, the rewards of shopping and consumption tend to be short-lived, but repetitive. There are those who believe that modern life in general favors a kind of widespread "low-level" addiction to all sorts of consumption, because the rewards are predictable, consistent, and frequent. As we learned in Chapters 2 and 3, this is the "wrong" reward schedule for how we evolved, and it makes us prone to the acceleration of consumption to get the same amount of reward value, similar to the development of tolerance for habit-forming drugs.[81] You have to do more of it to get a "reward dose" similar to what Mother Nature designed in the evolutionary world, when you stumbled onto an unexpected fruit tree when tired and hungry, got a rush of relief and satisfaction, and thereby learned to look for that fruit tree next time. Thus, when given access and means, the average human brain, perhaps particularly during youth, has to work hard to overcome the tendency to buy "stuff," and to forego those intense but fleeting rewards. Our modern life in wealthy, technologically connected societies encourages this perfect storm of circumstances, making behaviors that exacerbate our environmental quagmire easy, transiently rewarding, and more frequent than ever before in human history.

Rewards and Happiness—Are They the Same Thing?

We've seen that consumption is rewarding, but does it make us "happy" in a more existential way? This is a central point to understand, because happiness is what most of us say we really want most.

The science we've reviewed in prior chapters revealed the brain mechanisms needed for reward-based learning, which evolved over a very long time to help us survive and reproduce in the short term. We have seen how modern life in industrialized societies has altered what we experience and what kinds of rewards are available and prioritized, and how consumption of goods and other carbon-producing aspects of our lifestyles have accelerated as the Anthropocene has progressed. But is this the same as finding something rewarding in a more holistic sense? Is "reward" the same as "happiness"?

One of the main methods to study happiness is by using surveys, which ask people to rate their life satisfaction and then correlate their responses with circumstances, such as socioeconomic, cultural, and health status.[82] Longitudinal studies, some lasting many decades, attempt to discern what really matters in the long run. Such studies suggest life satisfaction is less about "stuff" and conventional "success" than it is about a sense of purpose and meaningful relationships.[83]

While the pursuit of happiness is as old as humans, contemporary scholarly studies of happiness also utilize neuroscience tools.[84] fMRI has been used to measure whether happier people respond to stimuli differently, and what parts of subjects' brains are activated when looking at pictures of happy or sad scenes, playing games that include social exclusion, or thinking about how far they have fallen from their ideal.[85]

As we learned in Chapter 2, resting-state MRI (rsMRI) has uncovered various brain networks that appear to be important for complex functions that require input from multiple brain regions. The most widely described is the so-called "default mode network," which seems to relate to the kinds of thoughts that occur while daydreaming, but also when reflecting on one's emotions, processing autobiographical memories, and thinking about the future. Exactly what these different networks mean and what purpose they serve is still the focus of intense research effort. Altered patterns of default mode connectivity have been noted in a number of mental health disorders, including depression and schizophrenia.

Some investigators have used rsMRI to interrogate psychological functions and traits in healthy people, such as happiness and resilience.[86] Additionally, the effects of interventions including intensive exercise and mindfulness training on the connectivity between different brain regions have been studied with rsMRI. After the intervention, improvements in objective measures of memory and cognition, as well as self-reported improvements in mood, self-esteem, and life satisfaction, have been correlated with concurrent changes in connectivity patterns between different brain regions. These changes have been interpreted as indicating that even adult brains have sufficient plasticity not only to improve focally to learn a new skill or task but also to acquire entirely new patterns that correlate with increases in life satisfaction—the measurement term for "happiness."[87] Such studies also suggest that the short-term rewards of purchasing and consumption have a different biological signature than do long-term life satisfaction and happiness. Thus, while the reward system figures prominently in decision making and behavior, critical during our evolutionary past to ensure we survived, it may not be our best neural guide to happiness. This will be worth keeping in mind as we think about changing behaviors in the upcoming steps of our journey.

Changing Brains and the Anthropocene

Our brains are amazing—fine tuned, capable, adaptable, to handle the incredibly complex tasks of human life in its infinite variety and with its infinite day-to-day challenges. They have been the source of spectacular artistic, scientific, and technological progress over a time frame that is a blink in the history of the earth.

But there is a mismatch between the pace of evolution of this extraordinarily able, pulsating, three-pound bundle of sparks and what we need to meet the challenge of this extraordinarily rapid Age of the Anthropocene. It is difficult for us to function in synchrony with our collective species-wide long-range needs to combat climate change and environmental decline. We are equipped with a brain reward system that simply didn't need to develop robust satiety—to food, to stuff, to stimulation—at the time it was designed and shaped by the way humans lived during the long millennial miles of the human journey before this last Anthropocene big toe on the last step. We do have brake pedals—we don't eat so much that we burst—but our brakes are relatively weak. Our excesses in contemporary life are fueled at least in part by unprecedented availability, the influence of youth on purchasing, and mass marketing that takes advantage of social changes in which media functions as a super-peer. Marketing also taps into the inherent gullability of humans of all ages to believe what they are told by sources viewed as authoritative. Consumption as a low-level addiction analogue in both youth and adults has been posited to reflect a mismatch between contemporary life, the potent rewards of immediate agency, and the infrequency of the biologically most effective reward schedule—varied, small, intermittent, unpredictable rewards. These mismatches may result in malaise, boredom and ennui, and pessimism, and a vague sense of not being happy, punctuated with intense moments of quite genuine satisfaction and even joy. But we simply do not have entrenched, old-growth, long-shaped powerful circuits that say, "Yuck! Too much stuff! Too much fossil fuel use! Too much unnaturally stimulating addictive entertainment! Alert, alert! Planetary problems ahead in upcoming decades! Ouch!! Pain!! Change course NOW!!"

But these are the brains we have, and these are the problems we face. We will need to use our advanced powers of observation, pattern recognition, analytic capacity, future prediction, goal-directed problem solving, social connection, and communication to bypass our deficit in reward circuitry input into this unprecedented problem. Yes, it's about survival, but our reward system hasn't caught up with teaching us, in this context, what we need to learn to survive. We'll have to dig deep into our reserve capacities, because this behavior change for survival won't happen on its own. First, we'll have to learn *which of our behaviors matter most* with respect to environmental impact, and what works best to change well-ingrained behavioral patterns at an individual and collective level. These are the subjects of the next stops on our journey. We can then determine how these principles and strategies apply to our environmental dilemma in the short time frame available for change.

6

Which Behaviors Matter Most

IMAGINE WALKING ON A PATH THROUGH WOODS with the goal of getting to shelter before nightfall. You come to a small river and have to decide whether to try to cross it. A fallen tree spans the water, a few rocks poke their heads above the turbulence, and the current swirls briskly. Cross or turn back? To make your decision, you rely on your senses—the sturdiness of the tree, the distance between the rocks, the whoosh of the current. You rely on your proprioception and sense of touch as you push on the tree to check its stability and springiness. Your memory provides knowledge of your own swimming strength should you fall in, the temperature of the air and water, and the available hours of daylight. Your threats and goals are visible, tangible, immediate, and recognizable as you make your decisions about which course of action you will take to meet your goal.

In contrast, next imagine a small gathering of people, one of whom is coughing, sneezing, and blowing his nose into a tissue. The virus he is likely carrying is invisible, but you've learned some facts about this kind of intangible threat, and you know to keep your distance and how to deftly sidestep shaking hands with the fellow with the hanky.

But climate change is a new and different kind of threat. Not only is it invisible like an infectious pathogen, but the cause-and-effect relationship for behavior choices are even more obscure than those you employ to combat the virus in the example above. A dizzying list of factors contribute, and each has its vehement champions as a focus of behavior change. It's coal, it's the cognitive dissonance of petroleum magnates, it's industrial agriculture, it's our meat-laden diets, it's methane, it's political corruption and poverty forcing people to cut down rain forests, it's our suburban sprawl, it's the downside and injustice of a capitalist economic system, it's the fact that people in developing countries who don't have them now will want refrigerators and automobiles, it's water, it's lots of things. The solutions also are myriad—we need changes in our individual choices, and major changes in society—infrastructure, economies, politics. No single cause or strategy alone will solve the problem.

For those of us living our lives in high-income industrialized countries, it's easy to get overwhelmed by all the things we know need to change to avert the threat, but we don't necessarily know how to make a difference. Our brains struggle with this kind of dilemma, and one reaction is to just throw up our hands and distract ourselves with the latest entertainment series or rearrange our proverbial sock drawer. To combat this tendency, this chapter examines how our behaviors affect our carbon dioxide output, as this is the single largest and probably best-studied part of the overall climate change equation.

As we think about behavior and carbon emissions from a brain point of view, it's worth remembering that the connection between our day-to-day actions and the carbon emissions that result from them poses a particular challenge to the way our nervous systems work. Carbon dioxide is not "pollution" as most of us conceive of it—dirty particles or directly toxic chemicals. We can't see, feel, taste, or smell it, and much of what falls in our per capita output column happens well outside our realm of direct experience. We are not designed to be distressed or rewarded by things we can't perceive, never mind measure. Understandably, then, most people have a vague and often inaccurate sense of which behaviors have the biggest impact on carbon emissions. Figuring it out takes homework, added on to other more immediately pressing demands of everyday life. We may be even less motivated to expend the effort because the potential personal consequences are less predictable than, for instance, the effects of an invisible respiratory virus, with which most of us have at least some familiarity. But there are researchers who study and measure the connection between behavioral choices and carbon emissions to help us learn which choices have the biggest impact.

Here it's worth addressing one additional dilemma in this arena. In research across all subjects, what is studied isn't purely determined by which questions are most important to answer. Instead, data may exist simply because a particular question could be answered with the methods and resources available. Data on individual people and their behavioral choices are available, because this is what's been in many ways easiest to study. Even the basic question of how much of the overall carbon emission burden arises from behavior at an individual level, compared to the degree to which more systemic elements are responsible, is challenging to answer. We will look at how some experts have tried to approach these thorny issues.

Is the Problem "Here" or "There"?

Carbon fluxes have existed as long as the planet has, but they were largely in equilibrium—carbon in vegetation ends up in the air and in the ground, goes through various cycles based on seasonal changes and other geochemical variations, and the carbon distribution in the oceans, air, and ground remained in balance for millions of years. When changes occurred, they tended to be very gradual—gradual enough that enough species could adapt through the process of natural selection for life to continue to evolve.[1] Thus, while there have been climactic changes and major extinctions before, they generally happened at a relatively slow rate. The critical difference we are facing now is that the carbon that was sequestered in the ground in the form of coal, crude oil, methane, and other fossil fuels is now being removed from equilibrium "unnaturally" for human combustion for energy, and released into the atmosphere at an unprecedentedly rapid rate. This is occurring so fast that the carbon dioxide and other greenhouse gases cannot be fully recaptured by the natural, established ebbs and flows that were predictable on "Earth 1"—the earth before the Anthropocene. This imbalance starts a chain of events that disrupts everything that biology had adapted to over millennia in unpredictable and potentially unsurvivable ways if left unchecked, because of the speed of its accumulation over a very short geological timescale—creating what writer Bill McKibben calls "Earth 2."[2] The steady rise in atmospheric carbon over the past 200 years and its resulting trapping of heat in the atmosphere like an invisible down comforter has occurred at a speed to which the biology of innumerable species simply isn't designed to adapt. These changes can be seen in many tangible ways, from human diseases carried by insects that are spreading into newer, warmer territories, to droughts that force mass migration and wars from conflicts for resources, to "once a century" record storms that happen year after year, to wind-whipped massive forest fires, to glacial shelves the size of whole states dropping into the ocean.

As noted in Chapter 1, among the high-income industrialized countries, the United States historically has been the main contributor of greenhouse gases to the atmosphere.[3] This is because for over 200 years, greenhouse gases have been the by-products of our large-scale industrial and agricultural progress. This rapid change was fueled by the promise and realization of many rewards, as we saw in Chapter 3, and until relatively recently we simply didn't have the means to recognize the invisible consequences. It's like a new drug that beautifully manages a crippling disease but after it's been on the market awhile somebody figures out that there's a serious side effect. But in the case of fossil fuel, it's a drug that people overwhelmingly perceived as improving their lives, and the side effect is usually first felt not by the user, but by people far away, in other parts of the world. And the causal relationship between the drug and the side effect is difficult to establish initially, as the effect can be caused by many other things. Because humans have never experienced something like this before, it's easy to both ignore and dismiss the connection. Not to mention that there are a lot of powerful people who are highly rewarded for maintaining the status quo.[4] This dynamic is a perfect setup for wishful thinking that fossil fuel use and environmental decline are not related, for deliberate dissembling, and for a long delay in recognizing the problem and taking action to address it.

In very recent times, China, a country whose enormous population, increasing industrialization, and expanding economy has steadily increased its total amount of CO_2 output per year, has surpassed the United States in annual total emissions.[5] But because the excess CO_2 concentration in the atmosphere lasts for many hundreds of years, more of the excess that currently exists has come from the United States than from anyplace else.[6] From the perspective of people living in United States and similar high-income countries, because our various types of waste per person are at the top of the global bar graph among countries, the problem to a significant degree is, in fact, about how we behave as individuals, if we are among the majority that consume these resources. Let's look at this issue in more detail.

Individual Behavior—Cost and Context

To understand the impact of individuals compared to businesses and companies, we need to get our numbers in order. Of note, we are focusing here on averages, which are weighted toward affluent people in high-income countries, as this is where most of the consumption and production of CO_2 arises. In this use of the word, "individual" or "consumer" refers to the person who purchases or uses the product or service for individual or household use.[7] It's important to note that an individual person's decisions can have an environmental impact far beyond what one does in one's "domestic," or personal, life; we will touch on that critical distinction in more detail later. But in the context of defining "individual" as a consumer for one's own household use, several analysts have attempted to parse out the relative proportions of CO_2 produced by individuals compared to that produced by institutions. This is by no means an easy task. Historically, most researchers have divided energy used and carbon dioxide produced into "sectors," although even what categories to include in the resultant pie chart varies among sources. One fairly typical breakdown includes "residential" (this is mainly where the "individual" role is counted), "industrial," "transportation," and "commercial."[8] But exactly what goes into each wedge of the pie can be debated. For instance, should transportation that people use to get to work go under "individual" or "transportation"? How about transportation to deliver goods that individuals will use in their households? The Environmental Protection Agency includes "Electricity" as its own "sector"—but it's used by people in almost every aspect of life.[9] The energy-dense health-care sector has a major environmental impact, as we will see in Chapter 9.[10] Often in research, categories are created simply by how the data are most easily obtained, by conventions in record-keeping used as information sources. It's not hard to understand that the way researchers "draw the line" around these sectors will have a major influence on their conclusions, and on how we figure out which behaviors in our daily lives matter the most. Do you add the carbon cost of manufacturing and transporting, say, a new washing machine to the consumer column, or do you put that carbon in the "industrial" column, pinning only the electricity and water use on the consumer? What about the energy and land use it takes to dispose of the washer once its useful life is over?

Some authors have attempted to apply to the "individual" column only those things that can be directly controlled by that person. Since you can't control the processes used in manufacturing, that environmental cost should go on industry, say these authors, and this approach leads to a percentage of around 40 percent of the carbon dioxide that can be traced back to those individual choices.[11] But wait, say other authors: the consumer can find out, and can make a choice based on the environmentally relevant practices of a given company. There are websites and apps that consumers can use, including the Environmental Working Group, Good Guide, and Done Good. These provide information on the so-called indirect aspects of goods and services—how much energy is used, carbon and other environmentally relevant substances are emitted, and so on, for the things you use in your daily life, before they even get to you, and after you're done with them. So, the individual who is consuming the end product should be responsible for the environmental profile of that product when a choice is available. When using this "consumer lifestyle approach" to attributing both direct and indirect energy usage and carbon dioxide production, the consumer activity ends up being responsible for more like 80 percent of energy utilization and CO_2 emitted in the United States.[12]

There is no consensus on exactly which metrics are the most accurate way to determine who is responsible for most of the energy utilization and carbon dioxide and other greenhouse gas emissions. But as a reasonable working figure for the purposes of our discussion, when you average across these various approaches, let's assume that companies and institutions are responsible for about half, and individuals in their private lives contribute roughly half of these emissions in the United States.[13] Even apart from what the exact percentages are, some researchers have argued that in the short term, changing individual behavior is the only effective means to reduce carbon emissions.[14] This is because longer-term solutions like new technologies and governmental policy changes, particularly if they involve complicated and often contentious agreements between countries, simply will take too long.[15] Thus, say these experts, what we do as individuals over the upcoming couple of decades is the most important bridge to slowing down emissions enough that technological advances in alternate energy production, new kinds of energy infrastructure, and implementation of effective national and international policies can "catch up" to the rapid pace of global warming. These are things we can do right now, without requiring the dismantling of an enormous petrochemical infrastructure and substituting new ways of doing almost everything in our lives. We have to act now, they say, and we have to act as individuals, and individual action has to spread among enough of us to matter. Other analysts focus less on change at the level of domestic decision making and more on infrastructure and incentive changes, especially those potentially already available.[16] Some argue that while change at an individual level might help, getting enough people to make changes that could make a difference isn't feasible.

For a problem of this magnitude, change at all levels is needed, and the different levels of course intersect. The individual consumer can't decide to buy an electric car if none exists within an affordable price range, or there are no charging station available at work. A company can't change its policy in a pro-environmental direction if the workers refuse to implement it or if the cost is prohibitive and destroys the business. Behavior change at the level of the brain will be needed for change to work at all levels, regardless of which gets accomplished first.

Additionally, an important neuroscience distinction needs to be made with respect to the very term "individual." There is a tendency to equate change of an individual's mind with change at the level of personal choices in the residential or household sphere. For instance, many people might say: "Individual behavior is the wrong place to look—it just won't have enough impact. We have to overhaul entire systems, like transportation." It's clear that changes at both levels are needed. But the changes from a brain point of view occur through the same mechanisms whether the consequences of specific behaviors and decisions affect just the individual making the change or affect many people. So let's look at big "systemic" change a bit further.

What has to happen to change a national transportation system? At a brain level, enough people in enough positions of authority to decide on and implement major changes must first have their minds changed about what to prioritize. They need to "mind" the climate crisis or its consequences enough that making the decision to change their behavior and implement meaningful action *is more rewarding than the alternatives.* Experts in social and political change, communications, the formation of movements, and cultural shifts focus on how ideas spread by social means over time, and ultimately reach enough people that values and behavioral norms within a society shift. These experts might study which individuals in what part of a political and economic decision-making apparatus need to change to effect the large systemic overhaul in question.

But at the neural level within each individual who is contributing to this shift, these specific changes occur the same way that all changes occur—when the millions of tiny weights of all the inputs we use to make decisions shift enough that a different choice is made. These weights might affect the decisions one makes about one's personal life, or they might affect the decisions one makes with respect to what political party to support, what legislation to vote for, what bill or referendum to push, what company to invest in, or what decision to support about operations at a large scale. The process is the same as what we saw in the "lunch break" episode in Chapter 2, when you decided when and how long to spend away from the office. It's the same mechanism people use to make a decision about whether to share or to hold on to money and whether to torture or revere cats that we explored in Chapter 3; the degree to which nature matters to individual people that we saw in Chapter 4; and the extent to which peer pressure and marketing play into preferences that we reviewed in Chapter 5. It may be influenced by the extent to which knowledge of the consequences of carbon emissions factor into individual people's decisions, and whether the people whose opinions they care about agree. In some instances, as we will see in Part 3, the carbon equation itself will play little or no role in the decision, which instead will be based on coincidental factors. The evaluative mechanism in the prefrontal cortex has to weigh in favor of a pro-environmental decision based on how rewarding that behavior is assessed to be at the moment a decision is made. If it's not rewarding in some way, people won't change. If it's rewarding to a large enough cohort of influential people who can effect change at a broader level in a society, then that particular systemic change will be actualized.

Thus, individual people may make decisions and personal choices affecting their own domestic lives; some also may make decisions that create change at the level of companies, policies, or laws. We will focus much of our discussion on consumer behavior, as this is easiest to quantify with respect to individual choice and carbon consequences. However, the same factors are at play for decisions made in people's other spheres of influence.

How *Much* Do We Have to Change?

Researchers have employed many different approaches to attempt to answer this question. The science of prediction of complex processes uses currently available data and involves a lot of highly educated best guesses. To answer the question of how much we need to change—that is, how much less CO_2 we need to emit—many researchers have relied on figures released at regular intervals from the Intergovernmental Panel on Climate Change (IPCC), an international body for assessing science related to climate change. This organization's mission is to provide governments and policymakers with "regular assessments of the scientific basis of climate change, its impacts and future risks, and options for adaptation and mitigation."[17] It does this by creating regularly updated reports based on the work of hundreds of scientific experts worldwide, peer reviewed and synthesized in a careful and standardized process by thousands of advisory scientists. The IPCC also serves as the advisory board to the United Nations Framework Convention on Climate Change (UNFCCC), which is a body made of representatives of 197 countries whose mission is to prevent dangerous human interference with the climate system. The UNFCCC sponsors regular climate-focused summits, including those at which major international treaties have been enacted, in Rio de Janeiro (1992), Kyoto (1997), and Paris (2016). You may recall that in 2007, the IPCC, along with former US Vice President Al Gore, received the Nobel Peace Prize for their work in identifying, assessing, and disseminating objective scientific information about climate change.

One critical benchmark provided by the IPCC during its years of work is an estimate of the number of degrees Celsius rise in average temperature, compared to pre-industrial levels, that policymakers should target as a maximal threshold in order to limit catastrophic consequences. The consequences are predicted to become catastrophic above certain average temperature levels because warming to a particular temperature is predicted to cause other changes that will further accelerate the rate of warming, at which point reversal of warming likely will be beyond our capabilities and the earth is unlikely to be able to continue to sustain many forms of life. As one common example of this kind of accelerating effect, temperature elevation leads to melting of polar ice, which leads to a rise in sea levels, which melts more ice. Since ice is white and reflects sunlight, and water is dark and absorbs sunlight, large bodies of ice converted to heat-absorbing water result in global temperatures rising even faster. This is an example of an accelerating, or "snowball," effect. There are many other predicted accelerators of warming; for instance, the arctic permafrost includes vast swaths of land that contain enormous amounts of frozen carbon-rich plants stored up over thousands of years. Because of global warming, this mass has begun to thaw and is releasing its enormous reservoir of stored carbon into the atmosphere. Thus, warming begets even more carbon, begetting even more warming, until the warming is unstoppable and can rise to levels that most life forms can't survive. This is not just because of the heat itself, but because of its consequences on all the interdependent webs on which healthy life depends. We can't solve the problem with more air conditioners, because to live, we also need fertile land to grow enough food, and all the complex interactive pieces of an ecological system needed to keep things alive, growing and healthy on Earth 2 and beyond.[18]

Data about these kinds of effects come from measuring changes happening during short intervals at the present time, comparing these to changes over similar intervals in the past, and then using these comparisons to analyze rates of change. Recent advances in technology have allowed an even longer view of atmospheric changes, extending back to the ancient past, by analyzing air droplets in samples from deep cores of ice dating back hundreds of thousands of years.[19] These kinds of data allow scientists to create mathematical models that predict what might happen in the future under different possible conditions—in particular, if additional CO_2 of varying amounts is added to the atmosphere. These different scenarios, called "Representative Concentration Pathways," or RCPs, model changes like temperature elevations, sea level rise, precipitation, ice melt, and other alterations over the different parts of the globe under conditions of varying concentrations of atmospheric CO_2.[20] Thus, the IPCC scientists have attempted to provide recommendations, or targets, for temperature elevations, below which this catastrophic accelerating scenario appears to be less likely, based on the available science.

The methods used and the specific data analyzed to create these temperature elevation targets are well beyond the understanding of most laypeople and are not without expert disagreement, which some critics have interpreted as a reason to distrust what they perceive as unfounded doomsday scenarios—or, alternatively, as too conservative.[21] However, for many in the scientific and policy communities, these continuously updated evidence reviews and target goals have served as the benchmarks for many of the research findings about what human behaviors need to change, and to what extent, to keep within particular temperature elevations that can be predicted to have specific consequences. Many scientific publications work within the framework that states that as a reasonable goal, we should aim to limit global warming to 2°C. This is a target that many believe could avert a "snowball effect" of warming that would be unstoppable.

So how do you translate that kind of overall goal to what people actually need to do? The answer requires complex data analysis and scenario modeling to try to fill in the puzzle pieces between global average temperature elevation goals and specific behaviors across the range of the enormous and diverse human population, but it is exactly what researchers in the field have been doing. To stay within these goals, taking into account that the population will continue to grow, it's estimated that we need to cut our global average CO_2 emissions per person per year from about 5 tons per year to 2.1 tons per year.[22] That means an even bigger drop will be needed for those of us at the high end of the range, here in the United States and in other carbon-hungry high-income industrialized countries that produce levels approaching 20 tons per year per person annually. We will need to cut our per capita emissions tenfold. This will require a potentially enormous changes in behavior. Where do we begin, and what does this mean for our lives?

Individuals and Carbon Dioxide

Two major strategies dominate most discussions about how to deal with the rapid rise of carbon dioxide in the atmosphere. Concentration depends on what amount goes into the atmosphere, as well as what comes out. In theory, then, carbon dioxide concentrations could be stabilized either by taking as much out of the atmosphere as humans put in, or by cutting way back on what human activities release—essentially to zero. These approaches are not mutually exclusive—according to the IPCC, they are both likely to be critical—but they may lead to different emphases on what kinds of behaviors and decisions need to occur. (In truth, the equation is not so simple, because carbon dioxide levels depend on many simultaneous interrelated fluxes of sources and sinks, some of which are already altered due to atmospheric, land, and ocean changes caused by human activities. But for this discussion, we will stick to a simplified formula.)

Let's first tackle the smaller part of the carbon-in, carbon-out equation. Can't we just take carbon dioxide out of the air and fix things that way? Since we're talking about behavior, what kinds of behaviors promote the removal of carbon dioxide from the atmosphere? Vegetation absorbs carbon dioxide from the air and stores the carbon in its wood and leaves as it grows. Thus, for individual people, reduction of atmospheric carbon dioxide can be accomplished by behaviors known to promote carbon absorption into plants, such as planting trees, facilitating reforestation, and restoring grasslands or other natural vegetative ecosystems. While plants absorb CO_2 during the day, they release some of it at night; but some is stored in the plant itself. When plants grow and even after they die, as long as they aren't burned or otherwise metabolized by organisms that release carbon as a by-product into the air, the carbon stays in the plant, and thus some carbon from the atmosphere is "sequestered," or captured, here on earth.

Can individual people make an impact through so-called "carbon capture" behaviors? Most people don't have individual control over the enormous areas of land that experts say would be necessary to offset the large amounts of CO_2 we've already added to the atmosphere in order to restore a balance; that would require more collective action. Land use issues make up an entire and important subspecialty of environmental science and policy. Using political and advocacy tools, much current effort by land-use environmentalists attempts to stem the rapid deforestation of the highly efficient carbon-absorbing ecosystems that already exist, but which are shrinking rapidly due to economic and population pressures. Deforestation creates a double whammy: it both releases carbon as trees are cut down and burned, and it reduces the amount of buffering capacity available within this uniquely dense and efficient photosynthetic system to absorb the carbon from the air and store it in leaves, stems, and ground.

A few organizations approach the task of reducing carbon dioxide by selling consumer products that are produced via replanting degraded lands rather than destroying existing rainforest for agriculture. Some also donate part of the profit to rainforest preservation or other restorative approaches. Getting something chocolate, and donating through your purchase to a group that is working to help the environment? Double reward! However, while such efforts in aggregate may be helpful, on an individual level, this has a small impact relative to other things individuals can do about their own contribution to carbon dioxide emissions.[23] And while the chocolate bar by no means is a bad thing (though life cycle analyses might question even that, as we will discuss), higher impact changes regarding diet and food choices are available, as we will explore below.[24]

Occasionally one hears advice along the lines of: "If everybody planted some trees, we'd solve the global warming problem." This might be rewarding—you get something new (hmm . . . river birch or red maple?) and pretty, and you've also done your part, so maybe you can check that box and don't have to do much else. But unfortunately, it's not that easy. As attractive as the idea is of going about what those in the climate change world call "business as usual" (abbreviated as BAU) and just offsetting our excesses by planting a tree, an analysis of the quantities involved shows that we can't even remotely solve the problem by planting trees alone.[25] That said, preserving existing trees, and re-forestation or restoration of other ecosystems such as undisturbed prairie, are additional behaviors in the right direction for stabilizing CO_2. You certainly can plant some trees—it's a good thing—but as you brush the dirt off your hands, don't let it falsely make you feel that you've done what you can do. There are other ways to help that will make a bigger difference.

Other approaches to actively lowering the accumulated carbon dioxide in the atmosphere involve developing and refining new engineering strategies to remove it via *carbon scrubbers* or capture technologies of various sorts.[26] Known as CCS, for "Carbon Capture and Storage" (or Sequestration), this approach requires energy to separate carbon by-products produced in power plants and to store them, most often underground in carbon-dense geologic formations. While some technologies exist to do this job, they are not yet in widespread use, and utilizing them adds to the cost of power generation. Nonetheless, carbon capture figures into many equations for how to deal with excess CO_2 in upcoming decades. Because this is easier to build into new power plants rather than to retrofit existing ones, it is estimated that curtailing consumption will still need to play the main role in achieving better carbon balance for the next few decades.

Geoengineering approaches under intense investigation include methods to counteract the warming effect of carbon dioxide by introducing into the atmosphere substances that might reflect rather than absorb sunlight, thus offsetting the warming effects of increasing CO_2. This approach has the potential to help stabilize temperature elevation (though it doesn't address other CO_2-related problems like ocean acidification). Unless you are a climate engineer, the behaviors needed to go in these directions will not be in your individual column—today. But they might be in the near future, when you will have to assess, decide, and act on your opinion about whether your government should take the gamble and proceed with these new, untested, scary, but possibly effective measures. What if something goes wrong, and we reflect too much sunlight, and create permanent winter? What if this works beautifully, and nature is "saved"? Your brain will use its various systems, with all their shifting, minuscule weightings of decision making, to help you evaluate the options. But you won't have much specifically matching experience to call on so that your brain can count on precedent to help with such a decision. As we learned in previous chapters, tilting your choices will be your inherent biological affinity for novelty versus familiarity, your comfort level with taking risks, your fear of negative consequences, your background reliance on facts and data, your overall status and situation in life at the moment, the opinions expressed by those around you and your opinions of those people, whether you think your opinion will make any difference at the voting booth or hearing, what happened to you this morning and last week and decades ago, what happened to your grandparents, and many other factors, some rational and some less so.

In his 2009 open-access book, *Sustainable Energy—Without the Hot Air,* Cambridge physics professor David MacKay helps get to the bottom of the question of what behaviors actually matter the most by relying on something fundamental: numbers. He points out that often, people who talk about energy, climate change, and behavior use words that don't help the listener decide what to do. For instance, knowing that our trash could fill a football field or that some large number of square miles of forest are cut down for development every week or that our cars spew X number of tons of carbon into the atmosphere doesn't help you decide which of your own behaviors to try to change, and often leaves people feeling bewildered and guilty without a meaningful course forward.[27] And when the suggestions offered by high school textbooks and government or environmental websites are compared to which behaviors actually have the most impact, these lists typically don't match.[28] For instance, recycling is on everybody's list, and it does help. But relative to other things you do in your daily life, it's a pretty small piece of the CO_2 equation (though it has other tangible benefits).

It's important to note that most people think that doing something "good for the planet" means *doing* without *something.*[29] This is known as "curtailment" in the lingo of behavioral strategy. Doing without, by and large, is not very rewarding. Sure, many people have at least a small part of their reward system, reinforced by temperament and experience, that feels good about "doing the right thing." But for most people, that reward isn't the main ingredient in most daily decisions. On the other hand, say a number of experts, you can choose to do things *differently,* rather than doing without, per se. This can lead to significant energy savings without making a major change in how you live. Oh sure, you might say—I can eat all I want, as long as it's kale and carrots. I can just forget about the good stuff—the cheeseburger and ice cream. That's not quite right, say these authors.[30] Because in the United States, we waste so much, simply stemming the waste may mean you get pretty much the same life, just without the hidden waste you didn't even know you were causing. We will get into how researchers examine how to best change behaviors in the next chapters, but for now, note that obviously not every behavior applies to every person. Some people don't drive. A few people live off the grid. And many people, even in wealthy countries like the United States, live in or at the edge of poverty, struggle just to get by, and have little freedom in their daily choices. So we are targeting this discussion to individuals with average incomes or above in wealthier countries, who have the capacity to make discretionary choices. If you want to know which behaviors add the most to your specific carbon profile, there are tools you can use, called "carbon calculators." These come in several versions, with most energy companies or the United States Department of Energy supplying them online. Just type "carbon calculator" into your search engine, and you can create your own profile and can learn what behaviors make the most difference for your specific situation by filling in a few blanks with relatively straightforward information. (Motivating people to take the time to do this is a separate issue, which we'll address when we discuss behavior change later.)

Right now, the biggest impact individuals can have in decreasing their own personal carbon emission volume is by lowering their energy consumption. Eighty percent of our energy comes from fossil fuels—carbon-rich compounds created from countless layers of dead vegetation sequestered and pressurized for millions of years underground. There's a lot of carbon—and energy—stored in fossil fuels. That's precisely why they work so well as fuels; they concentrate a lot of energy into small volumes. When people burn wood or other ground vegetation for heat and energy, it releases much less energy per volume than fossil fuels do, so that you just can't carry enough wood to fuel a plane. Thus, fossil fuels—coal, oil, gas—are the major source of human-generated greenhouse gases. This energy includes not just heating and cooling where you live and getting around from place to place, but the energy needed and carbon dioxide released from making, using, and disposing of all the "stuff" we consume. But here's another important fact: In the United States, not only do we use an enormous amount of energy, but we waste an enormous amount as well. This will be critical to understanding which individual choices have the biggest impact.

Currently, by a wide margin, the single largest contributor to human-generated carbon dioxide emissions is the burning of fossil fuels for energy. We use energy for all kinds of things—heat, electricity, transportation, manufacturing, food production. In his book, MacKay's central question is whether the United Kingdom could meet its energy needs using renewable, rather than carbon-based, sources; this would grind CO_2 emissions to a laudatory trajectory-changing halt. To answer the question, he calculates the quantity of the energy that could be generated by various renewable sources, stacked up against the quantity of energy consumed in various ways by the people who live there. While his focus is on helping people decide which policies on the renewable energy supply side of the equation make the most sense to support, we can use these same data to focus more on the demand side—how much energy people consume in industrialized countries. While renewable energy sources such as wind, solar, geothermal, and nuclear energy are a relatively small part of the power grid at present in most places, decreasing demand for energy might enable these sources to provide a larger proportion of the overall energy demand, thus significantly limiting carbon dioxide emissions. MacKay's estimates, along with those from other scientific and government sources, provide some useful information about the comparative impacts of various individual behaviors on CO_2 output.

On average across the entire world, MacKay estimates that we emit roughly 5.5 tons of CO_2 per year per person. In 2013, the World Bank estimated this number at 4.68 tons, quite a similar figure.[31] Whatever the exact amount, this per capita carbon emission isn't even close to being distributed evenly. In the United States, as mentioned previously, we average somewhere in the range of 20 tons per person per year, depending on the source of data and the dates of the measurements, while in India, the average is about one tenth that amount.[32] The differences among countries have been described as following a general formula introduced in the 1970s by scientists Paul Ehrlich and John Holdren, which represents the idea that impact on the environment (I) is a function of population (P) times affluence (A) times technology (T), an equation known as IPAT.[33] Our impact in the United States is so high in large part because, while our population isn't enormous compared to the world (roughly 330 million versus 8 billion), our level of affluence, reflected by our high gross domestic product (GDP), and our use of energy-demanding technologies in most aspects of daily lives, catapults us into the carbon-intense stratosphere of high-end users. Let's look at which activities use the most.

Ground Transportation

This is one of the single biggest ways an individual adds carbon dioxide to the atmosphere. Overall, transport of people and goods makes up about 20 percent of human-contributed atmospheric carbon dioxide. This amount is expected to continue to increase over time, both within the United States and worldwide.[34] Of the types of transport, road travel contributes by far the greatest percentage, several times greater compared to shipping, trains, and aviation.[35]

Collectively, getting from place to place by automobile is one of the most carbon-intense activities that people do. And we're doing more of it every day. Passenger car sales have more than doubled worldwide over the past 30 years, and in high-income countries, there is almost one car for every two people, compared to one car for every ten people in developing countries, although this ratio too is increasing rapidly.[36] So what's the biggest way to reduce its impact? Living a car-free life is one recommendation made by some analysts.[37] But this is the real world. For most people, alternatives for transportation currently lack the flexibility and convenience that auto travel provides.

Pooling data from a variety of sources, researchers have concluded that the single most effective way to reduce the carbon load of automobile transportation from where it is at present is to switch to a fuel-efficient vehicle. Carpooling to work, driving more slowly, keeping your car tuned up and tires inflated all help, but a fuel-efficient car—say, changing from a vehicle that gets twenty miles per gallon to one that achieves thirty or more miles per gallon—shaves off several times more than all the other options mentioned above.[38] The catch? You have to buy a new car! Some people make the point that if you buy a new car, your old gas guzzler still is out there on the road, and someone's driving it—so you've just added to the carbon load by buying your hybrid and increasing the total number of cars on the road. True, but gradually the older cars come off the road as they age, so as these are replaced over time by cars with better fuel efficiency, the overall carbon emission load decreases.[39] Our unit of measurement in this discussion is the carbon emissions attributable to each individual person—and you've decreased yours by substituting a fuel-efficient vehicle for your prior, less fuel-efficient, one. Currently there's a fair amount of choice, and you can get where you need to go pretty much the same as before—so you've done things *differently,* not done *without.* Substitution, not curtailment.

So let's talk about how people make decisions about purchases like cars. While your brain may get some reward from "doing a good thing for the environment," when it weighs all the little pros and cons of its daily onslaught of decisions, how might this one come out? Cars are a major purchase and an expensive one. Of all the motivations and rewards for spending that amount of money, the rationality-oriented reward of some relatively opaque "fuel efficiency," which is a tiny individual remedy for a big problem that's hard to even perceive day to day, may be a hurdle. The cost savings on fuel might be a carrot, but it's a relatively small one when you think about why people typically choose specific cars to purchase.

Years ago, I bought a sport utility vehicle that got twelve miles per gallon. My justification was that I spend a good part of my life taking care of people with terrible injuries, and I had young children of my own. The vehicle I chose was an early adopter of a steel cage construction and side airbags, it could seat seven people (cousins, in-laws), and could get through any weather if I had to get to work for an emergency. These were not terrible reasons, but it's also true that the vehicle company had a reputation for high quality, the forest green "rough and ready" look with roof racks appealed to my sense of being outdoorsy, and my old bedraggled "student debt" car with the drooping headliner held up by staples was made fun of by my colleagues. Nonrational reasons clearly played a big role in my decision. That car reinforced my desired self-image at the time—a responsible parent concerned about safety, an investor in quality, an employee with no excuses for not getting to work. This is of course the strategy that manufacturers and marketers utilize. Over time as my awareness of climate change grew, the obvious rational inconsistencies made me feel increasingly bad instead of good. As we saw in Chapter 2, something that was rewarding can become unrewarding or even aversive under changed circumstances, including additional cognitive input, even something as basic as food in extreme situations.[40] So while "big and heavy" won't likely be the characteristics of a high-mileage vehicle, if you could get other attractive features, and fuel efficiency—extra reward! So, what if fuel-efficient cars were sleek, or "cute," or had cool performance features? What if they were cheaper than gasoline-engine cars? For that matter, what if people like me could more easily change our perceptions, and recognize that something light and aerodynamic and with LED panels instead of heavy wood and metal and with a standard transmission is, in fact, more "outdoorsy" in the long run than a classic-model SUV? Then, as advertisers know, you might be more inclined to be rewarded by your choice. Auto manufacturers and sales staff will strive to convince prospective buyers that what they have to sell, which will vary depending on the regulations governing mileage standards, is what the target customer wants. They are not rewarded for fuel efficiency; they are rewarded for selling cars.

What is the impact of electric vehicles? They definitely cut down on point-of-use particulate emissions—other dirty things that come out of cars. The overall impact also depends on the manufacturing process (true for combustion cars as well), and where the car gets its charge. If you live in France, your electric car does great with carbon emissions compared to cars with combustion engines, because so much of the electricity there is generated with nonfossil fuel sources (mainly nuclear power).[41] In Holland, it's wind power that make electric cars a good partner for lowering transportation emissions.[42] In the United States, where a significant portion of the electricity grid is generated from coal-fired power plants, an electric car's advantage, carbon-dioxide-wise, compared to an efficient combustion engine car will widen as the use of coal in the electric grid decreases. (And it's already way better than a gas guzzler.) Coal, as we've learned previously, is a terrible fossil fuel for emitting carbon dioxide—it is significantly worse than oil, and far worse than natural gas.

Within the United States, some states use more coal than others. If you charge your electric car in California, you save lots of CO_2 compared to a conventional auto, but if in North Dakota, where 94 percent of the electricity historically has come from coal, your car has less of an advantage in its ultimate production of CO_2 than a combustion engine—though this clearly is changing with newer technologies and decreased use of coal over time.[43] Thus the expression, "switch to electricity, and green the grid" characterizes an overall strategy to decrease emissions from all aspects of energy use, from transportation to manufacturing to home heating, even before the grid has become predominantly renewable-energy based. Factoring the life cycle analysis of the manufacturing of an electric vehicle compared to a combustion engine car is also part of the equation.[44] But taking all these factors into account, more electric cars, especially those that don't require huge amounts of power for fancy performance profiles, powered by electric grids that contain a higher percentage of renewable energy from wind, solar, hydropower, and possibly biomass, will definitely make a meaningful dent in the amount of carbon dioxide generated from our daily automobile travel.[45] The amount of difference they make will change over time, depending on many factors having to do with improvements in efficiency and technology, the presence or absence of regulations mandating emissions or fuel efficiency standards, and electricity grid source changes in different geographical areas.[46] Consumers will need to make sense of a lot of conflicting information, some of which is provided by industries with a business-related agenda; to understand corporate decisions, one needs to follow the rewards.[47]

The important point here is that what you drive is one of the big ways you add to the global CO_2 complement, and it is one of the ways where individual choices can have the biggest impact.

Air Travel

Global air flight, like automobile transportation, is expanding rapidly, estimated at 6 percent per year on average, and increasingly environmental researchers emphasize that air travel has to be part of the equation when trying to figure out how to reduce CO_2 emissions.[48] Propelling something heavy into the air over long distances takes a lot of energy in the form of carbon-intense liquid petrochemical fuel. Airplanes add things other than CO_2 to the atmosphere, including increased cloud cover that also can increase warming.[49] There are some mitigators to decrease fuel consumption that the airlines themselves can take to improve fuel efficiency by flying different routes and reducing speeds, but making this economically attractive must balance the cost of extra hours of labor and other financial factors.[50] Fuels can make a difference, such as the switch from aviation gasoline to jet kerosene fuel in the 1980s.[51] Concentrated biofuel is an option, but it is still at an early stage of development. In order to make this option attractive to airlines, it has to be as cheap or cheaper than petrofuel, which will require further investment; additionally, the huge quantities involved may have other detrimental environmental effects.[52] Hydrogen fuels likewise are still in development and also may have other consequences; this transformative a change requires infrastructure overhauls and is likely to take decades. Thus, as in other aspects of the carbon equation we've discussed, in the short term, the most effective way to achieve CO_2 limits to keep us within the 2°C IPCC temperature goal will have to come from changes in the behavior and choices of consumers. At present, the main things you can do are to fly less, and participate in some carbon rebate programs, which we will discuss in more detail in Chapter 8. Voluntary individual limitations on air travel will not be welcomed by an airline industry that has been counting on an accelerating rate of growth, and it's fair to guess there might be major resistance and the creation of attractive incentives to keep people traveling by air. But some analysts argue that until alternatives are found, it's up to individual consumers to know about the environmental consequences of their actions and to make choices accordingly.[53]

As it turns out, of the choices that individuals can make, reducing air travel is one that has a considerable impact. Some writers have called it "possibly your biggest carbon sin"—especially if you live in a city and don't drive long distances.[54] Just one round-trip transcontinental or transoceanic flight adds 2–3 tons of carbon dioxide to the atmosphere per person. This is more than half your global average annual allotment of about 5 tons per person in the world at present. If you take several such trips a year, you'll get up to half of the already embarrassingly high US average emission volume per person of about 20 tons per year. And remember that to meet the goals to keep temperature elevation to 2°C, we have to cut this even further, to 2.1 tons per capita. For these reasons, almost everybody who has made a "list" of actions that individuals can do that matter the most with respect to carbon choices has included reducing air travel.[55]

But we have to travel for our jobs, to see the world, to understand what's going on, to meet people to share our knowledge and learn, you might say. Why, even researchers who study the effect of air travel on climate travel by air![56] Air travel is cheaper carbon-wise than other forms of travel that cover the same distance, say others. Yes, but you could cut out some of the less necessary big trips, and that would have a sizeable impact on your contribution for a given year, say some analysts.[57] This may require reorienting your rewards—substituting a different kind of business interaction or family reunion, for instance—as we will discuss when we talk about behavioral change in Part 3.

Housing

Heating and cooling your residence come in second only to transportation in how much energy most people use. So it's a big deal. Studies have shown that while turning down your thermostat, using less hot water, and setting your air conditioner slightly warmer may help, these curtailment strategies—"doing without"—have a lower impact than decreasing the waste inherent in the household systems most of us have where we live.

What behaviors have the biggest impact on lowering your individual CO_2 emissions depend in part on where you live and over which aspects of your living situation you have control. By and large, people living in cities have a lower carbon footprint than those in rural areas, mostly due to less need for longer distances of road transportation, and also because living in multifamily dwellings has some efficiencies of scale in terms of heating—you have to heat less if you only have one outside wall compared to four. About two-thirds of households in the United States are owned by the occupants. There is a huge amount of waste in heating buildings in general; by some estimates, at least 30 percent of heating energy is completely wasted.[58] This is true for many households—it is estimated that 80 percent of older homes are underinsulated.[59] If you are a homeowner in a cooler climate, especially if your home is older, weatherization of your home will have the greatest effect—insulation in the attic, double- or triple-pane windows that hold in the heat—because most of your energy use is for heating, and a great deal of the energy is lost through the cracks in your roof and around and through your windows. Pooling data from many countries, it's been estimated that retrofits that make homes more energy efficient could cut heating and cooling energy usage by 50 percent or more.[60] In the United States, an analysis of weatherization measures suggests that on average these can cut energy utilization by at least 20 percent. Homeowners can do even more by purchasing energy-efficient heating and cooling systems, appliances, and lighting, as we will see in Chapter 8.[61] Also now in the energy appliance lineup: your computer system. While this was thought to be insignificant in the past, the overall carbon footprint of computer systems is now equal to other household appliances in many homes.[62]

As global temperatures rise, more of the energy load will be for cooling rather than heating, but making buildings more energy efficient still will save carbon and dollars under those scenarios.[63] Actions that many of us think we should take—turning off the lights when we leave the room, turning down the thermostat—make a difference, but to a much lower magnitude. This isn't to say they aren't helpful. Some things have small impacts, but if they're relatively easy and enough people do them, they can add up to a lot on the level of a power grid—like changing your light bulbs. We'll talk more about collective and individual action in the next section, on changing behavior. But if the goal is to make the biggest dent in your own personal consumption and its resultant CO_2 emission volume, knowing what behaviors have what impact—and what are the barriers and incentives to create those changes—are critical things to understand.

Food

Agriculture and food production are responsible for at least 25 percent of all greenhouse gas emissions.[64] Perhaps because food is on people's minds several times daily (as opposed to, say, attic insulation), this topic seems to have captured the public's interest and imagination more than some drier aspects of climate science. Writers like Michael Pollan (*The Omnivore's Dilemma; Food Rules*) and Colin Beavan have popularized the idea that locally grown food has a more favorable environmental and health profile, because, among other features, it requires less shipping, may be fresher, and is often produced using fewer pesticides or other chemicals.[65] Pollan is famous for making it simple: "Eat (real) food, not too much, mostly plants," and "Don't eat anything your grandparents wouldn't have recognized as food." Large-scale, so-called "industrial" farming practices and processed food manufacturing have come under fire in recent decades for being both dangerous for the environment and for health. Arguments about genetically modified food crops that resist certain types of pests or pesticides or can grow in specific conditions, thus increasing yields but with potential downstream environmental (and, say some, health) consequences, remain rampant. The global rise of diets containing more processed foods, refined sugars and fats, and meats, with deleterious effects on both the environment and health, has been termed the "diet-environment-health trilemma," and is considered one of the cornerstone targets for addressing global warming and other aspects of environmental decline.[66]

An enormous amount of research has gone into this area. Much of it supports the idea that a locally grown, sustainably produced, mostly plant-based diet has advantages for minimizing environmental decline resulting from agriculture, such as the effects of fertilizers, large-scale water requirements, and the production of greenhouse gases including methane. Beef production in particular has a high environmental and carbon emission profile, as well as being blamed as a contributor to a number of health conditions when consumed in quantities typical in a US diet.[67] Beef and lamb have a greenhouse gas emission profile that is 250 times that of legumes for the equivalent gram of protein.[68]

As in the study of the environmental effects of transportation, housing, or consumer goods, the concept of life cycle analysis—following everything that happens, from seed to table to scraps—increasingly has been applied to food.[69] This approach highlights that food production, processing, and transport involve complex processes. The differences between locally produced, small scale, organic food and much of our contemporary, conventionally produced food involve a multitude of steps and systems, and the environmental and health advantages are not always aligned neatly with one or the other.[70] Not all vegetarian diets are healthy; in poor environments with limited food diversity, meat may provide essential nutrients not otherwise available in that setting.[71] Thus controversy arises at every step, and a lot of the controversy ends up in forums and media sources that laypeople—all of whom eat—may encounter.

How are ordinary people to wade through this messy landscape? One approach has been an increasing advocacy for utilization of labels that denote not just the nutritional value of food, but also the environmental profile needed for its production and transport. Like an Energy Star rating for a refrigerator, food labels could provide the buyer with a comparison of, say, apples grown in the next town compared to those grown across the country, or of different lunch options in a cafeteria.[72] This information is not currently readily available; for instance, most people don't realize that rice is associated with five times more greenhouse gas emission compared to an equivalent amount of protein in wheat. Other aspects of nutrition are important as well; some foods have high greenhouse gas profiles but provide needed nutrients, can be good for health, and have a more modest environmental impact if consumed in small quantities.[73] Thus, exactly how these metrics would be created and vetted are the source of ongoing discussions, with many different stakeholders involved.

Population

Some researchers and policymakers have tried to help people by creating lists of the most important, impactful things individual people can do to reduce their carbon footprints. Most of them avoid the topic of population size—even though it's an obvious multiplier of greenhouse gases and everything else relevant to the question at hand. But Seth Wynes and Kimberly Nicholas, writing in the journal *Environmental Research Letters* in 2017, created their own list of the four most impactful things people can do, and they tackle this issue head on.[74] By their calculations, having one fewer child for those living in high-income industrialized countries can be projected to decrease CO_2 emissions by a whopping 58.6 metric tons per year—an order of magnitude greater than any other behavioral choice. They don't suggest having *no* children—just choosing to have one fewer. Yet they point out that this is virtually never mentioned in science textbook sections on climate change, nor in government informational documents or recommendations. No carbon calculator includes this option as a variable. While one can postulate all the political and social reasons this might be the case, Wynes and Nicholas's figures certainly provide food for thought.

Which Behaviors Matter Most

So now that we know what individual consumer behaviors have the biggest impact on carbon dioxide output and climate change—switching to a fuel-efficient or electric car (or doing without one altogether), improving the energy efficiency of your home, eating a mostly plant-based diet, taking one less plane trip per year . . . and maybe having one less child—what do we do with this information? How is this knowledge converted into action, keeping in mind what we know about how reward factors into behavioral choice? And it should be emphasized that parallel choices having environmental consequences are being made by individual people in their other roles in life every day—making decisions about their companies, institutions, investments, policies, and laws—that may have vastly broader impacts that those we make that affect our own personal spheres. Getting ourselves—and getting other people—to make changes is hard, though some changes are harder than others. We next need to learn some of the science about changing behavior overall, and behavior relevant to environmental preservation specifically, as our journey continues.

Part 3

Changing the Brain

7

Behaviors That Are Easy and Hard to Change

WE'VE NOW LOOKED AT HOW OUR BRAINS evolved with a reward system designed for survival advantage and have considered which of our individual behaviors make the most difference in contributing to our environmental crisis. We've seen how our neural equipment is working in a new world compared to when it evolved, and thus is being called upon to help us adapt to novel and rapidly changing circumstances. So if our behavior needs to change, how hard might that be? Is it even feasible, in the time scale in which change has to occur?

Diverse fields including education, psychiatry, psychology, public health, business, advertising, and marketing all use behavior change as currency. Researchers have uncovered general principles about which kinds of behaviors are easy or more difficult to change, and what strategies are most effective, often using insights from neuroscience. These general observations about behavior change in different contexts will be important to recognize, so that we will be able to apply them more specifically to environment-related behavior as we pull together all we've learned and begin to round out our journey.

If we are so adaptable, why is behavior change hard at all?

When we are talking about changing behavior, often we are talking about changing *habits*—things you do repeatedly the same way most of the time, without giving them much thought. Some behavioral scientists define habits more precisely than the usual conversational form of the word, and they differentiate habits from goal-directed behavior, depending on whether the behavior changes if the value of its outcome varies.[1] But other scientists frame habits as a more flexible part of a continuum of behavior—you have learned to do something the same way repeatedly, so that it becomes practically automatic. You are able to do things differently if the circumstances change, though it can be a slow process, requiring a shift to different brain circuitry for a reset.[2] For present purposes, we refer to people's customary way of doing things as a typical behavior that occurs reliably in response to a particular stimulus or circumstance, most often due to extensive experience of achieving their goals in the same way. You walk in the door to home or work, you hang your coat in a particular place, and your routine is pretty well set—where you put your keys, what you do next, and so on. When facing a decision, you respond in a characteristic way—you have learned, by trial and error, by being shaped by authority and your own experience, to have an opinion and preferences that tend to guide your behavior in particular ways in response to particular scenarios. What's going on in your brain?

As you may recall from Chapter 2, these customary, typically repeated behaviors are designed on purpose, deliberately created by the specific way the brain forms the network connections responsible for them, to be strong patterns that are resistant to change. As we have seen, this is true for virtually all creatures, from sea slugs and fruit flies through humans. Overlearned behaviors are designed to become entrenched in your neuronal circuitry after a considerable period of learning through trial and error, and thus are based on experience in the world. They typically serve you well, and changing them carries some risk.[3] Imagine how inefficient you would be if every decision required you to start from scratch—where should I hang my coat? Where should I put my keys? We are creatures of habit by evolutionary design, because this is what worked best to help us survive. This is why changing entrenched behaviors when new evidence comes along to suggest they are not ideal—especially if that "not ideal" reason isn't immediate, with direct, here-and-now consequences—is not an easy task. We are working against the grain when we try to do so.

We are pretty good at changing our typical behaviors when our immediate circumstances change. When we move to a new place, after a period of adaptation, we regain the ability to automatically remember the new place to hang our coat without thinking about it, and after a while, the new grocery store feels pretty normal. As with other human traits, some of us are better at adapting to this kind of circumstantial change than others, and akin to different people liking novelty versus familiarity, this difference also likely played a role in our collective survival evolutionarily. It makes sense that if most of us couldn't cope with the addition of a new sibling to the family, taking on new responsibilities with maturation, turnover in group leadership, separation from loved ones, changing capacities and insights with aging, and other ordinary life changes that force us to change our habits, we'd be in a sorry state. Conversely, if some people weren't particularly attached to how things have always been, we might not have the skills in the group to save and pass on historical cultural knowledge, which also is important to our success.

While generally we can adapt and learn new responses, it's harder for us to change our habits when it's not because our circumstances have changed, but just because we receive new *information*. Changing behavior requires a change in rewards. The old way must become relatively less rewarding, and the new way must offer more potent rewards (or a chance to avert something bad). Thus doing things in a new way is especially challenging if the purported reason to change behavior is not something readily perceived with our own senses. Parables, literature, and scripture are full of tales of people who didn't heed warnings from others about things they couldn't see themselves, from Aesop to Doubting Thomas to Little Red Riding Hood. These reflect a truth about brain design: if the information describes a theoretical or future-based risk that is not an immediate threat to the recipient, it generally registers less compellingly. Information that comes from a source we don't know personally and with whom we don't have a trusting relationship also undermines the acceptance of the new information, especially if there is any uncertainty about its veracity.[4] Deliberate disinformation spread for political or financial goals further impedes change. These barriers are compounded if the behavior you're being encouraged to stop doing *is itself rewarding*. That bacon you've eaten since childhood is now bad for you? Are they sure about that? It tastes great, and we've always had it, so how bad can it be? It's protein! Wasn't it just last week they were saying eggs were bad for you? And white bread? And orange juice? Heck, there goes breakfast! Who are these "experts," anyway? We're all going to die of something, so pass me some more of that bacon, thank you very much.

In light of this known resistance to change, what evidence can we call upon to learn which approaches have worked best to change difficult behaviors, and in what contexts? Before we focus specifically on behaviors influencing climate change in Chapter 8, examples of research on positive reinforcement and substitution, culture change and social learning, cognitive dissonance, and nudging will help us understand some of the best-studied general principles for facilitating particular types of difficult behavior change.

Positive Works Better Than Negative

Our nervous systems are designed to respond to and learn from both positive (rewarding) and negative (aversive) stimuli, as we saw earlier. This makes sense, as we need mechanisms to react effectively to both attractive and dangerous things we encounter. By studying hungry rats and pigeons who were taught to press a lever or peck at a target for food pellets, in the 1950s, psychologist B. F. Skinner and colleagues worked out in painstaking detail what schedules of reward (giving food pellets for some, but not all, of the lever presses or target pecks) were most effective at encouraging the animal to learn a new behavior. (As an interesting aside, this whole line of inquiry arose because the pellets had to be made individually using a pill machine, and it simply took too much time to make enough pellets to provide one for each lever press! Thus the science of behavior was permanently changed because of something entirely coincidental.)[5]

This type of animal experiment showed that small, intermittent, variable rewards are most effective at facilitating behavioral persistence and learning new tasks—similar to the evolutionarily conditions in the real world.[6] If at first you don't succeed, try, try again. If you can't find the berries, or reach the fruit, or chip the spear point—keep trying, and sooner or later, you'll achieve your goal.

These studies also shed light on how behaviors are unlearned ("extinguished") when no longer paired with reward—the rat presses the bar but no pellet comes out. If we couldn't unlearn associations, we'd spend our lives in futile behaviors that don't lead to reward. However, this doesn't address the problem of changing behavior that continues to remain firmly associated with a reward, at least in the short term—but may be associated with a remote or hypothetical future risk. This is the situation for climate change, and we will examine this situation in more detail shortly.

It's clear that different personalities respond differently to positive and negative reinforcement.[7] But as a general rule, reward works better than punishment for both new learning and for changing habits.[8] In education, positive reinforcement has been used successfully to eliminate disruptive behavior in the classroom and in one-on-one interactions with challenging learners.[9] In these circumstances, the undesirable behaviors were rewarding—they garnered the student attention or resulted in avoiding an unpleasant or difficult task. But immediate positive reward for desirable behaviors is most effective at eliminating the problem behaviors. Not surprisingly, it also has been shown that caregivers generally prefer using positive compared to negative interventions—people prefer praising to punishing their charges when given the tools to do so.[10]

There are many other real-life contexts besides education in which positive reinforcement has been shown to change behavior more effectively than do negative consequences. Getting hospital workers to wash their hands before and after every patient encounter, essential for reducing infection transmission, is most effective when participants are rewarded for doing so rather than punished for lapses.[11] In the public health realm, rewards for changing behaviors that lead to decreased risk for health complications, like eating healthier food or getting more exercise, work better than just telling people to change.[12] Whole companies have arisen to create devices that reward you with emoticons or other positive feedback when you take enough steps or walk enough distance. In industry, changing behaviors to achieve goals, such as implementation of safety protocols in the workplace, is accomplished most effectively by utilizing positive reinforcement.[13] In businesses, changing corporate culture to make workers engaged and feeling rewarded and loyal to the company is thought to be most dependent on praise, recognition, and personal positive reinforcement from superiors in response to specific jobs done well or goals met.[14]

How does this work in your brain? There are distinct but overlapping neural systems to deal with reward and punishment, and interestingly, we seem to be more sensitive to increasing amounts of reward than to increasing amounts of punishment.[15] Studies on both animals and humans have shown that learning is enhanced in relation to the amount of positive feedback and reinforcement received. These studies also show that connections between the motor system involved in enacting the behavior, the prefrontal cortex involved in decision making, and reward circuitry all are strengthened during learning that is paired with positive reinforcement.[16] Thus adding reward to the equation helps us learn new information, but it also facilitates unlearning habits and substituting new behaviors—that is, creating behavioral change.

Some behaviors are notoriously difficult to change; these often involve strong biological drives and resistant habits. Let's take two of the most difficult—overeating and addiction.

One thing is clear: just giving people information doesn't work to change these behaviors. These are examples of very entrenched patterns—there are rewards associated with the behavior, but the behaviors persist even when negative consequences also become associated with the activity. Often it's the case that the reward is strong, immediate and directly connected to a persistent or habitual behavior, while the negative consequences are more remote and obscure. In overeating, the rewards are immediate and basic; the downsides are remote, gradual, come from authority figures with whom one may not have a personal relationship, and may not even be certain ("Nobody in my family has diabetes. I might not even get it!"). With addiction, the behavior persists even though the reward becomes less and less potent over time, and the negative consequences become more and more predominant. This is why many scientists describe addiction as resulting from "hijacked" reward circuitry. As we saw in Chapter 2, the system isn't designed to work with such substances added to the equation; they disrupt the fine-tuned balance provided by nature. This happens most frequently with drug addictions, but it can happen with other addictions as well—to eating, gambling, shopping, and video gaming.

So what does work for these very difficult problems? Here, too, the principle of immediate positive reward seems to be key. Not long-term reward ("You'll feel better in six months"), not a theoretical reward ("People may like you better"), but rewards that substitute something immediate for the reward that the individual is passing up. In many cases, the main tool is social reward—establishing a social support system that regularly provides reinforcement and encouragement. Guiding people to set small short-term goals and then rewarding them when these are met is an important strategy—this taps also into the reward of agency, or control over what you can make happen, as we saw in Chapter 3.[17] Approaching complex behaviors from a number of directions at once is helpful (changing eating habits as well as adding exercise, for example).[18] For the addict, social rewards, substitution rather than curtailment (e.g., nicotine patches, methadone, healthy food instead of unhealthy food), dissociation from contexts that trigger habitual responses (being around other people who use drugs, or the ice cream aisle in the supermarket), and tackling several interrelated problem behaviors at once seems to increase effectiveness.[19] Many of these multidimensional approaches are those used in well-known weight loss and "twelve step"-type addiction treatment programs. The biology of this kind of change is starting to be understood: it reflects that ultimately, the reward system is malleable—something that wasn't rewarding before can become rewarding and can substitute for the old reward. As an extreme example, we saw in Chapter 2 that fMRI scans of patients with anorexia show that something as antithetical to basic survival as one can define—not eating—can activate the reward system.[20] As we saw with the medieval cats in Chapter 3, with a very big dose of external influence from trusted authorities weighting the input from all those billions of far-flung synapses, what we find rewarding can be altered, even sometimes when it's against our actual best interest.

Here's one example that applies the principle of substitution of "good" rewards for destructive ones. By itself, this approach is likely to fail without additional social and economic support. But it's a good example of a creative strategy to target one powerful reward with another, a veritable Wonder Woman–supervillain showdown. While superficially it may sound simple, it aligns with what we have learned about brain rewards in prior chapters. The Lullaby Project is a program that teams at-risk mothers-to-be, like those with opioid addiction, with musicians, in order to create lullabies for the babies they are expecting.[21] The women, with guidance, write poems to their future children, and the musicians help transform the poem into a song. The emotional wallop of music—think soaring strings, arranged by empathetic artists, set to your own words—combined with the deep well of parental protectiveness and care that may be as basic a part of biology as anything people can experience, are further enhanced by hormonal changes designed to strengthen those emotional bonds. It's a perfect combination for a potently rewarding tool for behavior change. It's also a reward the women report they can go back to again and again, their own personal resistance anthem. The power of this approach is amplified by the well-recognized strategy of what happens to people when they make a public statement and receive affirmation from a community of supporters; the women have performed their lullabies in places as public and credible as Carnegie Hall.[22] Writing lullabies won't cure opioid addiction by itself. But for an individual who can tap into other powerful neurobiological rewards as a motivation to change, accompanied by a change in physical context away from the habitual haunts that trigger addictive substance cravings and replaced by the context of caregiving and parenting—something shared and generally supported by an entire society—it's a brilliant bridge strategy for behavior change.

This example serves to show how one strategy for behavior change substitutes one set of strong rewards for another. There are as yet only scant and preliminary data for how well this particular approach works—that would require long-term research (and funding) comparing this approach to others over many years—an arduous task.[23] However, it's clear that even with intensive support, relapse is the rule more than the exception in the particularly difficult behavior change areas of weight loss and addiction. Both carrot and stick approaches struggle to overcome deeply entrenched, unnaturally overdriven brain circuitry, even when the behaviors it directs are extremely maladaptive to the individual and to society.[24]

For this reason, some researchers believe that such behaviors must be directed by societal interventions. The ban on transfats and taxes on sugary beverages are examples of government-level interventions to curb obesity. Health warnings and advertising bans on tobacco are others. Whether addictive substances should be illegal is an argument way older than Prohibition. So the lesson from specialists in this kind of behavior change is that these are difficult problems, they require multimodality treatment, including social support and other positive reinforcing substitutions, but they still fail often. Thus, say some, enforced societal solutions—decreased availability, criminal sanctions—may be needed. But even societal-scale solutions depend on changes in the priorities, viewpoints, and choices of enough individuals to successfully create and enact the new rule. The decision to change happens in the brain, by the mechanisms we have learned, whether you're the legislator signing the bill or the store owner choosing the product to sell or the customer deciding whether it's worth the price with the new tax added.

Some Behavior Changes Require Culture Change

As we learned in Chapter 3, we are built to be "gullible," in the sense of believing what we are told by authority figures, as part of our species' evolved survival strategy. Who we view as an authority depends on our specific circumstances, including community influences, exposure to different ideas, and trust. Some changes in behavior require overcoming well-established beliefs about what kinds of actions lead to what consequences, what sources of information are trustworthy or suspect, and what kinds of behavior are right or wrong. When these ideas are shared by a group of people, these can be termed "cultural beliefs" or "cultural values."

One striking example of attempts to change behaviors that conflicted with entrenched social values comes from the Ebola epidemic in West Africa in 2014–2016. The virus—highly contagious and highly lethal—was spread by human-to-human contact during illness and from contact with the bodies of people who had died from the disease. Rapid spread was compounded by well-entrenched caregiving and funeral practices that evolved under the influence of cultural, economic, religious, and political factors to play a critical role in the social order of the people living in the region. Some of these practices developed to protect against the extractivist history of the region in colonial times, affording people in the community some degree of hard-won local control over their affairs.[25] This particular Ebola outbreak ultimately sickened over 28,000 people and killed more than 11,000. As it was spreading rapidly, something had to be done to change behavior—and in this case, change was a life-or-death emergency.

During the early part of the epidemic, the mainstay of public health efforts included radio announcements and media communications with the goals of encouraging people to do the following: 1) isolate affected individuals and bring them for treatment provided at centralized medical facilities (Ebola Treatment Units); 2) avoid touching possibly affected individuals; and 3) desist from funeral practices involving ritual washing or touching the body.[26]

As a general rule, in the West African population affected by the epidemic, sickness and death were viewed as a reflection of things the ill individual did wrong, typically by behaving against social norms. Alternatively, sickness might be a consequence of a curse by someone else, who might be living or dead.[27] One common cause of dead ancestors acting out to the detriment of a family was an improper burial that did not include settling debts or other aspects of appropriate funerary traditions. To add to the challenges faced by workers trying to confine the epidemic, within the community there was justifiable distrust of local officials based on unfortunate past political experience. Additionally, rumor and misinformation played a role. Some community members believed that Ebola was being deliberately introduced into the population by rivals or other malevolent outsiders, that health facilities were stealing valuable blood or organs from sick patients who died, or that other ulterior motives justified distrust of the intent and outcomes of government authorities and medical clinics. For one thing, wherever "invading" health-care workers went, disease seemed to follow—so what's to say they weren't spreading the disease? As the epidemic progressed and capacities to deal with it were overwhelmed, calls to hotlines that authorities encouraged people to make when Ebola was suspected went unanswered, desperately sick patients were turned away from full clinics, and bodies were left uncollected by trained workers. These events did not help the cause of overcoming distrust and widespread fear.[28] *People have reasons for their behavior.*

But gradually, as community members saw firsthand that health-care workers also succumbed to the disease, and that some Ebola victims brought for early treatment did survive, conspiracy theories declined. Especially in the urban areas, where a higher percentage of the population had some formal education, a degree of trust already had been established between community members and the health-care workers who provided medical care for obstetrical and other conditions. Thus there was a degree of openness to the possibility that some of what was being touted as ways to prevent spread of the disease might be effective. The most successful strategies for behavior change utilized the principle of "cultural humility," along with recognizing the resources of strength that already existed within communities, in order to approach, engage, and partner with respected community leaders. This partnership approach was used to design strategies that would respect local customs and needs while still achieving goals based on the scientific concepts of infection control and prevention.[29]

Public health workers aided by anthropologists enjoined community religious leaders to sanction changes in religious funeral practices, which helped the community members accept the recommended changes—watching funerals from a safe distance, cremating remains, and having special prayers at the service that substituted in function for some of the more standard rituals.[30] Such social learning measures helped overcome the extreme reluctance of family members to fail to follow established procedures to help the dead successfully transition to the afterlife and thereby incur wrath that could plague the family for generations.

But what about the "don't touch a sick person" rule? Imagine being a mother with a child who comes down with signs of possible Ebola. The child develops fever, bloody vomiting, diarrhea, and dehydration. At their outset, the symptoms could reflect any number of common illnesses endemic to the region. How is the mother supposed to care for and comfort the child without touching? Is she supposed to walk away and leave the child to suffer alone? Yes, she may know intellectually that if this is Ebola she could get sick and leave the rest of the family worse off without her if she dies. But not touch the sick child? This goes against every inborn and socially reinforced behavior of a caregiver in almost every culture.

This is an example of where behavior change has to be realistic. As in the case of overeating, you can't just tell people "don't touch." You can't curtail; you have to substitute. As one caregiver said, "It will be impossible that my child or husband is sick and I refuse to touch them. I do not have the courage or heart to do that."[31]

Instead, people in Ebola-stricken areas wanted advice on how to care for family members while also decreasing their chance of catching or spreading the disease. Working along with respected community leaders, both men and women, health workers found ways to change behavior that would decrease disease transmission without asking caregivers not to tend to and touch sick family members. There weren't nearly enough gloves and fluid-resistant gowns even for the hospitals with known Ebola cases, never mind to hand out to the public. Instead, campaigns showing community members how to use plastic bags and raincoats to protect themselves from contagious body fluids provided a way to change behavior without abandoning people's basic sense of compassion and duty to care for dependents. People donned what they could find and carried sick relatives to the clinics to avoid contaminating taxis or other people. While the epidemic was tragic, it was, ultimately, successfully contained. Rebuilding provided its own social and economic challenges—caring for orphans and reintegrating survivors (whom some viewed with suspicion) into the community. Change has to fit the culture, while at the same time encouraging the culture to change from within to meet a new goal, informed by new knowledge.

Cognitive Dissonance, Denial, and Behavior Change

Crises like pandemics can cause obvious conflicts in motivations, but conflicts happen in everyday life as well. Because we are socially motivated, most people are strongly rewarded by approval, by the good opinions of others, and may engage in behaviors that accrue these rewards at many social levels.[32] But in some circumstances, this can lead to so-called "cognitive dissonance." This occurs when our beliefs conflict with our behaviors, so rather than changing our behaviors to match our beliefs, we change our beliefs to match our behaviors. The classic example of cognitive dissonance is smoking. Nicotine is one of the most addictive drugs known, and when a conflict arises because of information designed to change behavior (public health announcements, warnings on the package label), many people find it easier to change their beliefs ("The science isn't settled on that—this is an overstatement" or "My grandfather smoked and he lived to be 92") than to change their behavior to align with available facts.

Sometimes people become so overwhelmed by the magnitude or complexity of change needed that they just give up and withdraw—this is another form of coping mechanism, and it may be characterized by other psychological descriptors, like denial and compartmentalization. People may change their beliefs to fit their behavior ("It's not really a problem") or decide to tackle unrelated problems they feel they have a chance of solving. The profusion of nail salons in many cities may reflect the general truth that weight loss and exercise are so difficult and require such major changes in behavior—but getting your nails done makes you feel like you look better, and doesn't make many demands. This is the often-subconscious tactic of giving in to a pressing urge to reorder your sock drawer right when the whole house needs to be cleaned, or rearranging your colored pencils when the big project is due tomorrow.

With respect to behavior change and social approval, cognitive dissonance can occur when someone's work or professional role conflicts with values or beliefs that person holds outside of the work sphere. Some authors cast this kind of compartmentalization as one of those dilemmas particularly common in modern industrialized societies. If you work for a company that engages in unfair business practices that undercut the competition, but you are rewarded when these actions benefit the company, you may decide that those practices are quasi-legal and that's just the way successful businesses have to operate to get ahead in the real world. If you win Employee of the Month and your picture is up on the company poster in the break room, it's hard to walk away just because their practices and your sense of what's right and wrong have a minor conflict. Many people can live with cognitive dissonance until something external forces a change; bad corporate behavior may cause a public outcry that hurts business, or new regulations may close loopholes. And reflecting the neural diversity among people, if the employee happens to have the trait of finding rule-following and "outing" rule-breakers highly rewarding, is more swayed by allegiance to a moral authority than to the company leadership, and at the same time is not completely averse to the risk of "sticking one's neck out," cognitive dissonance may find a different solution—a whistle-blower may emerge!

Nudging

In their 2008 book, *Nudge: Improving Decisions about Health, Wealth, and Happiness,* economist Richard Thaler and legal scholar Cass Sunstein make the point that behavior change can be guided by "choice architects" who can affect people's decisions by strategically arranging the context in which their choices are made.[33] These theories are based on decades of research by many investigators in fields we have encountered previously, namely, behavioral economics and psychology. They arise from the recognition that much about the choices we make is not rational but instead is influenced by all kinds of seemingly extemporaneous factors. While the psychological principles are not new, proponents recommend using this knowledge of human behavior to change choices in predictable ways as a matter of policy.[34] Thaler, Sunstein, and other nudging advocates suggest that relatively small changes in how choices are presented can easily tip the scales to influence people to choose things that are best for their health and well-being, without eliminating their full range of choices. This approach follows a principle called "libertarian paternalism" (or, by some authors, "asymmetric paternalism")—wherein the framing of choices to make some more attractive or easier than others is based on what people themselves agree is "good for them."[35] "Nudges" typically influence choices at an unconscious level. A classic example of nudging is arranging the food in a school cafeteria to encourage kids to choose the healthier options by putting the fruits and vegetables at eye level, while the junk food, though still available, is harder to find. Another example is using "opt out" programs as the default for employee savings plans rather than "opt in" plans, in order to make it more automatic—and thus easier—for people to save money for retirement. Similar "opt out" policies have increased the percentage of drivers who note on their license that they are willing to be considered an organ donor. The idea here is that framing choices to make the "better" choice the easier one doesn't compel people to make a particular choice—it just makes it more likely that they will do so. This makes nudging as a behavior-change method more acceptable to those who object on principle to authoritarian, choice-constraining regulations that determine for people what's best for them.

Nudging has gained a lot of attention as a mechanism for changing behavior, and has considerable appeal because it doesn't eliminate choices, but helps to subtly influence people, at low cost and effort, to adopt particular behaviors.

There are many well-studied examples of how arranging the order and relative availability of choices has been shown to influence people's behavior in significant ways. Advertisers and marketers have known this forever; just look at the array of impulse-buy items artfully arranged at eye level at the checkout counter, not to mention "buy two, get one free" campaigns. More recently, theorists have described multiple types and classifications of nudges, based on whether choices are influenced by changing the default option, changing the physical environment of choices, whether nudges are transparent or overt, change something about the stimulus, or depend on other techniques.[36] Nudging has been used successfully without adding any specific communication of new information at all. Examples from widely divergent applications include improving road safety by visual cues—optical illusions, more or less—that unconsciously influence drivers to slow down on a curve, and increasing the frequency of handwashing in rural Bangladesh by painting bright footprints to guide schoolchildren to the handwashing station after using the latrine.[37] Nudging has been studied extensively for its effect on food-related decisions, in settings including supermarkets, cafeterias, schools, and recreational venues, and has been shown to make a significant difference in what people choose.[38]

The concept has its critics, however. Some have argued that nudging doesn't really work or that it is authoritarian, conceptually fuzzy, and unethical, in part because it manipulates people's unconscious choices.[39] Creation of governmental policy behavioral advisory teams in the United States, Great Britain, Canada, Australia, Denmark, and other countries has prompted some commentators to describe them as nanny-state or Orwellian tactics, incompatible with the basic premises of informed democracy, since behavior can be manipulated without people's awareness and without the opportunity for them to weigh in or discuss reflectively how choices are being influenced.[40] The counterargument is that choices are influenced by the context in which they are presented anyway—it can't be avoided—and so why not make the context one in which the "healthier" or "wiser" choices are easier to make? Proponents make the case that market forces and advertising very frequently work against health by nudging behavior in an unhealthy direction, which health-focused nudging simply counteracts.[41]

Measuring the long-term effectiveness of instituting nudges to achieve health goals in large groups of people has been more challenging, because of the varied determinants of complex behaviors, and because most studies don't follow subjects' choices or health outcomes long-term.[42] Culture may play a big role as well. As one outlier attracting attention in the public health world, Japan has a low obesity rate despite being a highly industrialized modern nation. But the reasons may reflect not simple nudging so much as the homogeneous and culturally entrenched values regarding health and nutrition compared to the overwhelming free choice in cultures like the United States. In Japan, school lunch is mandatory, uniform, and carefully planned by nutritionists. Lunchtime is integrated into the pedagogical experience to teach traditional Japanese culture, nutrition, hygiene, and manners rather than to relax and socialize as a "break" from the school day. Companies in Japan typically mandate annual health screenings and provide detailed follow-up and support for workers who begin to show signs of health issues such as being overweight. Companies invest in this effort because it saves health-care costs in the long run. In contrast, as a stand-alone strategy for combating obesity, simply adding a little nudging may be insufficient in a more heterogeneous, free-choice environment with few uniform societal norms, such as the United States.[43]

Despite some controversy, nudging strategies do appear to change behavior. But the degree to which changes are long-lasting or have health benefits over time remains incompletely studied.[44] Nonetheless, nudging as a behavior change strategy has gained attention as a means to shift behaviors of larger numbers of people in more widespread ways as they interact with institutional modifications.

We now have a broad view of what kinds of general behaviors are more amenable or more resistant to change, and how the design and function of the reward system play a major role in behavioral decision making and choice. We also have a sense of what kinds of strategies and approaches work best to effect changes in behavior in a variety of contexts. We are now equipped for the important next stop on our journey—to examine how behavior change strategies, including those discussed above, have been applied to behaviors specifically relevant to our environment crisis.

8

Strategies for Pro-Environmental Shifts

PICTURE NOW THAT WE'VE PAUSED ON OUR JOURNEY, catching our breath at the crest of a hill with a view forward. We've had to work some to get here! We've learned some neuroscience and neuroanatomy, delved into the workings of the reward system's role in decision making, evaluated the biophilia hypothesis, studied the brain function / modern life / accelerating consumption interaction, analyzed which behaviors add most to carbon emissions, and acquired some knowledge of tools used for difficult behavior change. Now, finally, we're peering ahead from our "savannah view" overlook, straight into our future with increasing climate change. How do we use what we've learned to move forward?

In this chapter, we'll survey what strategies have been tried in order to change attitudes and behavior specifically relevant to climate change and environmental decline. We will start to assess whether the knowledge we've gained thus far on our journey, along with what others who have worked in this arena have learned, helps us more effectively change our behavior and move us on to a better path.

The first step toward effective behavior change is to understand that *the nature of the problem is different* from a tangible challenge like how best to cross a river, or even "invisible threat" challenges like how to curtail a pandemic. In the scope of problems one might tackle, climate change falls into another category altogether. Let's next look at how this has come to be understood, and what strategies have evolved to respond.

From Simple to "Wicked"

Approaches for researching and changing behavior relevant to the environment have evolved in parallel with the scientific understanding of the nature and scope of the problem. Rachel Carson's classic environmental book *Silent Spring* was published in 1962, and people could see for themselves things like smog, trash, urban decay, polluted waterways, and wildlife decline.[1] Biologist Paul Ehrlich's *The Population Bomb* was published in 1968.[2] The idea for the first Earth Day in 1970 came from Wisconsin senator Gaylord Nelson in response to witnessing a giant oil spill in Santa Barbara, California. There was a growing recognition that humans were increasing in number and impact and were damaging the planet. The hope of the Earth Day organizers was that a designated time for largely youth-led demonstrations and events might harness the energy seen in other contemporary public movements, like those for peace and civil rights, toward environmental concerns. Ultimately such an effort might galvanize a public response leading to policy change. With some bipartisan support and organizational energy from Harvard student Denis Hayes, Earth Day was initiated as the beginning of a movement to bring environmental awareness to the general public and to government and industry leaders. As part of this successful tide, the Environmental Protection Agency was formed in December 1970, and legislation including the Clean Air Act and Clean Water Act mandated major changes toward environmental improvement by law.

At that time, in the United States, some of the major environmental concerns were litter, air and water pollution, outstripping limited resources including food and energy, and population growth. Global warming and climate change had not yet reached widespread recognition as an invisible threat equal to or greater than the sensorially obvious problems like blight and industrial waste. The field of environmental psychology had been created, journals were founded for publishing research results, and psychologists began to study how to facilitate behavior change in earnest.[3]

During the next two decades, many studies were undertaken regarding environmental behavior. Not surprisingly, as we saw in other realms of public health in Chapter 7, the preponderance of evidence suggested that increased knowledge correlated only slightly or not at all with more environmentally sound behavior at the level of an individual person. However, techniques borrowed from other contexts of behavior change that we saw in Chapter 7 were shown during these early years to be successfully applied to this sphere as well, targeting behaviors such as use of public transportation, reducing energy use, and recycling. Thus, socially reinforced incentives to promote behavior change, including the use of direct reward, competition, goal setting, and publicly stated behavioral commitments, along with involving known community organizers and establishing group norms for particular behaviors, were demonstrated to be more effective than education alone.[4] However, as the scope of the environmental crisis and the contextual transience of the behavior changes elicited in pilot experiments became clear, it was apparent that additional approaches were needed.

The next several decades brought the distressing realization that indeed, we now were fully within the Anthropocene era, and that climate change from greenhouse gas accumulation was the central and most critical environmental threat to the entire planet and all its interconnected ecosystems. Addressing the problem of accumulating greenhouse gases was made more difficult by its very invisibility; pollution you can see; trash you can trip over. This was no straightforward problem; in fact, it was becoming undeniable that changing the most important environmentally relevant behavior was a "grand challenge" or so-called "wicked problem."[5]

Such challenges result not from a single event or one-dimensional problem, but from intersecting trends and changes, which in turn influence each other, certainly an apt way to characterize global warming and environmental decline. It also embodies a social dilemma—one that occurs when what's disadvantageous to the group is favorable to the individual. With respect to climate change, high-carbon lifestyles are the root of the problem for the group, but on an individual basis are advantageous and convenient—another version of the tragedy of the commons.[6] There are enormous financial, technological, and political factors pushing with great force to maintain the status quo. This grand challenge was going be even harder than curtailing smoking, an intransigent but more focused behavior. This was not a moon shot, where at the end of a shared outlay of extraordinary team effort, you have a visible victory with someone planting a flag to exultant cheers and hugs and the mission is accomplished. No, instead it became clear that addressing climate change would require alteration of multiple behaviors on the part of multiple people in multiple different contexts, over a long period of time, often requiring total shifts in what might be considered "rewarding" up against long-engrained behavioral habits and decision-making conceptual frameworks. It would mean trying to effect changes in behavior that many people weren't even sure was necessary, for a problem they couldn't viscerally understand, explained by people they didn't necessarily trust based on things they weren't sure were true, and with no prior experience to suggest otherwise. It would require change at the level of individuals, institutions, economies, and governments around the world, often requiring self-sacrifice, or at least changes in priorities and routine ways of doing things, on all these levels for a common good across boundaries of all kinds. It would mean those with resources would have to change on behalf of those without, and the desire to change would be met with powerful resistance on many fronts. New methodologies to accommodate the complexity of the behaviors being studied would need to be developed for a problem whose nature and scope had never before tested human ingenuity.[7]

Values, culture, economic considerations, and politics typically both cause and are needed to solve "grand challenges." They require a panoply of approaches and novel strategies rather than a single solution. Moreover, problems of this nature are not typically "solved" per se, but hopefully at least managed until further scientific and social changes evolve to help address them, such as technological innovations. Both short-term, temporary "bridge" changes—the kind many theorists believe are the most critical and pressing at the current time to prevent the snowball effect of acceleration of warming—and more long-range systemic changes are needed for this kind of challenge.[8] Interventions often must be instituted on insufficient data because the problem exists in uncharted territory. But these difficulties are not reasons to fail to act—a tempting avoidance response for difficult problems that seem overwhelming, in the context of accelerating—and often disquieting—cultural and technologic change.[9] And as we saw in Chapter 5, avoidance is made easier by our modern life in the developed world, with its constant availability of habit-inducing entertaining distraction that is eager, for a small fee, to take our minds off more vexing long-range behavior change that we can't even fully understand. It's all so contradictory, we don't even know what to do, even if we wanted to.[10]

Why Prevention Is Harder Than Treatment

An additional barrier to climate change action is that humans generally have an easier time treating a problem than preventing it. Consider the relative reluctance with which most people approach preventive health measures like flu shots or colonoscopies, compared to the intense urge many of us feel to do something, right now, if we develop a new symptom or discover a lump. This tendency arises from the way the brain works, and this evolutionary shaping has relevance to strategies for changing environmental behavior. In our 40-day walk from San Francisco to New York that we started in Chapter 1, representing the relative time scale of evolutionary events from the formation of the earth to the present, we first encounter early humans at the edge of Times Square. We are 223 yards and two and a half minutes from our final step, marking the end of our journey at the present time. As we have noted, just the length of our big toe on the very last step of our timeline represents the duration of the industrial era of the Anthropocene.

For early humans, neural mechanisms to perceive immediate threat were well established—predators, storms, snakes, and spiders. But our nervous systems didn't yet have mechanisms to understand or communicate remote future disasters. We didn't have knowledge of invisible threats—infectious disease, toxic substances, the long-range effects of our own behavior. Even if a community member could conceive of a distant, intangible threat, there wasn't the means to readily validate or share the information. And what is the reward value of avoiding a potential, largely invisible threat? It's also dependent on cognitive processing, a very secondary reward compared to, say, a good meal when you're hungry. You might feel a little surge of altruistic reward, or of agency by averting catastrophe, of "being right," as we saw in Chapter 3. But when the problem avoided is so theoretical and remote, it's much less of a tangible reward than successfully constructing a protective garment or weapon or building a shelter to stave off threats you know well through your own direct experience, and have direct consequences to you as an individual.[11]

Climate change and environmental decline present particular challenges for behavior change, in part because both the phenomenon and the "fix" are many steps removed from our immediate sensory perception and the evolved workings of our reward biology. Unequivocally, climate change is here already, but for many people in the high-income world, it's hiding in plain sight. When it's perceived at all, it's generally seen by most people as something we have to prevent rather than to treat, and this alone can lower its prioritization. While this perception may be changing because of the increasing frequency of events causally related to environmental decline, we aren't physiologically equipped with carbon dioxide sensors, and most of us don't have strong personal experience of direct threatening effects that can be unequivocally, viscerally attributed to climate change. Rather, we know about them second hand from often impersonal sources, and even threats that affect us more directly—storms, droughts, heat waves, fires—can be ascribed to natural or external causes rather than perceived as causally linked to our own personal behavior.

As an additional problem for our behavior change motivation, "limiting global warming to 2°C" just doesn't sound like that big a deal, or like something that should be so difficult. That's because a 2° temperature fluctuation is something we all experience regularly in our day-to-day lives, and being 2 degrees warmer for those who live in a temperate zone just doesn't seem, well, all that bad. Comprehending the deleterious consequences of this average temperature rise at a global level takes work, concentration, background knowledge—it's homework, and for what short-term reward? It just makes you feel bad.[12] Is it the least bit surprising people would rather spend their mental energies on more immediate positively rewarding things and might be distrustful of those who say this is such an awful problem? Most of us simply don't fully comprehend why 2 degrees will change things so drastically—this is something with which we have no direct familiarity or experience—and at some level may be skeptical about the whole thing. Additionally, if you are being rewarded in other spheres of your life for behavior that runs contrary to environmental goals, cognitive dissonance is likely to kick in, and it may be easier to change your beliefs to fit your behavior than to deal with resolving the troubling conflict.

Let's say you've actually experienced flood and wind damage from a hurricane, and you believe this might be related to climate change. Even so, your logical goal-directed behavioral response might be to get a generator, a stronger sump pump, and wind-resistant shutters. That's how you prevent a problem—get prepared, so that next time is less bad. But changing to a fuel-efficient car? Planning one less plane trip next year? Becoming a vegetarian? Intellectually, you might know that this is what may help in the "big picture." But your brain is likely to feel pretty distrustful that this is the best way to protect yourself in the short term, and the reward is likely to be pretty weak—it just feels like a totally insignificant drop in a very big ocean. As for social rewards, you might even worry about setting yourself up for ridicule more than praise. And after the hurricane, will you vote for the candidate who supports wind power tax breaks, or the one who got the damage in your neighborhood fixed up promptly? Your brain has an easier time creating allegiance to the one who did something you can perceive as tangible, immediate, and personal.

As we have seen, our brains were well designed to learn associations and behaviors for overt survival-related goals. They were not subject to evolutionary pressure to perceive or react to the kind of threat exemplified by environmental decline—one we learn about mostly through information from people we don't know and communicated through language, a relatively recent development in our collective history, superimposed on a much, much older reward system design.[13]

Whose Behavior Needs to Change?

Assume that the Intergovernmental Panel on Climate Change calculations are essentially correct, which predict that a business-as-usual approach to human activity has a high chance of leading to critically dangerous levels of climate warming over the upcoming decades, and that drastically decreasing carbon dioxide release into the atmosphere during the first half of the twenty-first century is the mainstay of stabilizing this trend. Furthermore let's assume as correct the evidence that currently, the developed world is the main contributor to greenhouse gas release, and that the average per capita carbon output (nearly 20 tons per person per year in the United States) is approximately ten times the target required to stabilize temperature at the desired 2°C elevation (approximately 2 tons per person per year; see Chapter 6).[14] Finally, as outlined previously, let's assume that roughly half of the carbon emitted in high-income industrialized countries reflects activities of individual people in their daily lives, and half is due to things beyond direct individual control. We clearly need to change, but where do we begin?

Unlike other, more familiar problems, with climate change it's not been universally clear who should do the changing. Most societies in the developed world have someone whose job it is to respond to most of the major threats we've seen before. There are agencies and infrastructure for dealing with diseases, for natural disaster events, for war. Though new versions may arise, we can all generally agree that these are threats; we can easily create a mental image of their potential consequences and agree that something needs to be done. And we are generally willing to do our part when directed to do so by people we recognize as authorities (even, sometimes, in the context of misinformation).

But whose job is climate change "prevention"? Some theorists have divided the context of behavioral change relevant to the environment into domains in which people act, including private (at home), public, organizational, or activist.[15] Research has shown that different sociopsychological and sociodemographic characteristics are associated with how people behave as individual consumers, environmental citizens (such as belonging to advocacy groups), and supporters of policy.[16] Others have divided spheres of change into *macro* (policy and economy), *meso* (industry, corporations), and *micro* (individual consumer) levels.[17] Finally, behavior that has environmental impact can classified as direct (e.g., cutting down trees) or indirect (making policies that influence environmentally relevant behavior). All of these theories, constructs, and classifications arise from a variety of disciplines and viewpoints, adding to the complexity of studying and analyzing what actually works to solve the problem. In fact, intersection between all these levels must occur to promote policy changes that make an effective difference in behavior at the scale needed.[18]

As we saw in Chapter 6, roughly half of the contribution to carbon output and environmental change in developed countries can be attributed to choices made by individuals in their domestic lives. Some analysts contend that action at this "micro" level is the fastest way to effect change as a critical bridge until technology, industry, infrastructure, and institutions catch up, because change at these larger-scale levels takes much longer.[19] In addition, adopting yet-to-be-developed novel technologies, once they are refined and commercialized, ultimately also requires change on the part of individual people. So, say these analysts, change at the level of individuals—the micro level—is the best place to start.

It's clear that changes at the level of an individual person make a small contribution to the big problem; some experts think focusing on individual rather than global approaches is misguided. But the aggregate of many people making similar changes can be considerable. Other analysts have pointed out that some behavior changes required for climate mitigation involve the so-called "social dilemma"—the action is a relative negative for the individual, but a positive for the big picture challenge. For instance, turning down the thermostat may be good for energy conservation, but who wants to feel chilly at home?[20] While many mitigating behavior changes are not, in fact, negative for the individual, one element not often directly addressed in research is the following: whose job is it to try to create behavior change in those within this individual-action segment who don't change on their own? You may be worried enough to factor high mileage or battery operation into your car choice, but what about your neighbor with the combustion-based SUV, who uses it for commuting and general transportation, and who thinks the whole premise of climate change is bunk? To deal with that wide variability within the populace, who are the people that need to change their own behavior in order to become change agents for other people? And whose job is it to provide the direction and the incentive? This is part of what makes climate change a "wicked" problem.

Some thoughtful analyses suggest that we have the technology and know-how, right now, to do what needs to be done, by utilizing some relatively straightforward "wedges"—ways to slow down carbon accumulation—that could provide the margin we need to keep climate under control for the next 50 years, *if we would just do them*. This would provide the critical bridge until new non-carbon-based energy sources are developed and implemented. For instance, if we cut energy waste with means we already know, increased average vehicle mileage with technologies that already exist, decarbonized energy production in ways we already are doing but could do more, captured carbon with technologies we already have but don't use fully, and stopped further deforestation, we'd hold things steady.[21] But this optimistic view still depends on policy shifts, regulations, cooperation among levels of organizations and governments, incentives for businesses, and engagement and cooperation of essentially the entire general public. In the long run, many have argued, such changes likely would be far cheaper than the foreseeable effects of climate change as it continues to worsen. However, until economic models factor in the cost of paying for the societal, economic, and health effects of climate change in the future, thus creating a significant financial incentive for behavior change, figuring out who and how people will "step up" remains challenging.[22]

Some people think it's the job of concerned scientists and advocates, policy makers, and government leaders to change the behavior of the general public.[23] Policy changes, regulations favoring environmental goals, and funding for scientific research come largely from government agencies and legislators, and the priority given to these goals vary with the views of individual office holders, and how environmental issues rank against competing priorities. To try to predict what factors might influence decision making in this sphere, game theory techniques used to study individual economic behaviors (see Chapter 3) have been applied to modeling potential behaviors of institutions and governments negotiating climate change policies.[24] The fact that some countries have enacted more pro-environmental policies and their associated behavior changes than others is not because their brain equipment is different, but because the social and political incentives provide enough weight to influence people's decision making and available options at many layers of society.

Are Legislators the Ones Who Should Change?

One major challenge noted by policy analysts studying democratic societies is that elected officials are motivated primarily by being elected, but election cycles are short. Political considerations have played major roles in some environmental landmarks; for instance, viewing a political rival on television addressing enormous crowds in New York City on the first Earth Day in 1970 influenced Richard Nixon to establish the Environmental Protection Agency, to steal some of that thunder.[25] This is not to belie the fact that individuals who go into politics may do so with sincere belief in important causes that they intend to carry forward. However, policies that may inflict "pain" on voters in the short term for long-term societal gain do not enhance the chances of politicians keeping their jobs. This makes advocacy for climate change mitigation policies, which may engender unwelcome change for some constituents and campaign donors, a difficult choice for many politicians around the globe.[26] Neither voters nor politicians are particularly well equipped brain-wise for long-term problem-solving lasting decades or centuries, nor do most democratic political systems favor long-term priorities over short-term concerns. *Prevention is harder for most people than treatment.*

Voters tend to choose their political candidates on more immediate issues, primarily on economic, security, and social policies, though a high proportion acknowledge concern about climate change when surveyed.[27] In Chapter 3, we noted the neurobiological and upbringing-related traits generally categorized as "liberal" and "conservative" in personal and political behaviors, and saw that climate change belief and advocacy are more widely aligned with liberal political platforms worldwide.[28] Of interest, consistent with the principal of cognitive dissonance, the very act of voting for the party with a weak or absent environmental platform may flip voters' view in the direction of climate change skepticism after the election, thus aligning beliefs with behavior, presumably to resolve tension between the two.[29] Other analysts studying the interaction between voter behavior and environmental decline predict that the increasingly tangible effects of climate change may lead to increased officeholder turnover. This is because constituents tend to vote out incumbents when their personal circumstances decline, as is likely to occur more frequently as climate change progresses and its direct and indirect effects—weather events, economic fallout, migration of climate refugees, food and water supply problems, health consequences—increasingly are felt closer to home.[30] As more voters prioritize environmental issues, more politicians likely will do so as well.

Thus, support of elected officials who promote environmental goals requires an informed populace that shares these priorities. However, consensus is more challenging when scientific information is actively resisted and undermined. Regarding government support for environmental initiatives, some researchers have suggested that more local or regional approaches can work better than trying to wait for consensus at the national or international level, which is more difficult to attain.[31] For political candidates, environmental platforms are more successful if accompanied by proposed mitigation strategies that can be associated with some short-term tangible gains (for example, more jobs, economic stability, energy savings, more green space, cleaner air and water), what some have termed the "No Regrets Approach." Pro-environmental measures that benefit constituents and don't require them to change much are those most likely to garner voter support, even when they are recognized by the voters themselves to be less effective at combating the problem. Additionally, more local or regional approaches matched to the specific circumstances of that region may work better than trying to find something to which a broad swath of the population will agree, which is more complex. For instance, living closer to the coast and being more highly educated tend to increase support for environmental policies.[32] Lower on voter favorability are measures that might be more effective but require more controversial or difficult changes on the part of individuals (such as paying carbon taxes or major financial penalties for automobile travel), especially if they are content with the way things are in their own lives.[33] As we have learned, well-engrained choices are designed in the brain to be resistant to change, especially if the behavior to be changed is associated with a pretty comfortable life at the present time, and the threat is perceived as comparatively remote, or as someone else's responsibility.

How an environmentally relevant political message is framed can influence voter acceptance. Framing is the emphasis of one attribute of an entity compared to its opposite; on average, people will pay more for a burger labeled 75 percent lean, emphasizing the positive, compared to the same thing labeled 25 percent fat.[34] In the environmental realm, many economists have noted that a carbon tax, in which the costs of environmental mitigation are factored into the cost of fossil fuel or other carbon-producing activity, would be one of the most effective ways to limit carbon emissions. This is because the added cost would motivate individuals to change their behavior, and also would provide incentive for companies and governing bodies to find alternative energy solutions. Presently, while not mandatory, so-called "carbon offsets" are available for environmentally concerned consumers for many goods and services. These are additional costs that customers can contribute when purchasing things like airline tickets, with the additional voluntary payment being used for carbon capture or alternative fuel. So does calling it a "tax" versus an "offset" make a difference to whether people would be willing to pay the extra cost? In one survey of almost 900 people contacted through the internet, when environmental costs were labeled a "tax," self-identified Republicans and Independents tended to be unwilling to pay the extra price, compared to when the same costs were labeled as an "offset." On the other hand, Democrats tended to respond the same regardless of how the cost was labeled.[35]

To make the kind of progress this crisis demands, steadily more legislators have to change. In Chapter 2, when we first explored how the brain makes decisions, we encountered a legislator about to make a decision on a climate bill. You now know that how this person votes will be determined by innumerable factors, conscious and unconscious, relevant and irrelevant. He or she will be making a gamble. All of the reward circuitry—the SNc and VTA, the prefrontal cortex and nucleus accumbens, the anterior cingulate and limbic system all will play a role. Representing literally millions of swirling events, on balance, at the moment of the decision, will the complex evaluative circuitry decide that the reward from voting yes outweighs the reward from voting no? Thousands of factors—the varied views of constituents, implications for reelection, campaign promises and funding, the consequences of prior votes on this topic, a personal sense of what's the "right" thing to do, the views of one's family members, a recent protest event, whether the person who introduced the bill is a friend or competitor, a news article read this morning, today's weather, a recent doctor visit, a movie seen this week . . . each influences the decision in tiny ways. Change happens when citizens, business leaders, campaign donors and volunteers, experts, writers, activists, colleagues, and political leaders concerned about climate change influence legislators in many ways and at many scales by ultimately making yes more rewarding than no.

Corporate Social Responsibility

What about change originating in industries and companies, whose short-term economic self-interest may not align with necessary alterations in business-as-usual behavior? Decisions in these realms have a large influence on the "other half" of the environmental problem, outside the control of individuals in their private lives. Business leaders get their rewards primarily from their corporate boards and shareholders, largely reflecting their profit margins. Managers get their rewards from their supervisors. Hundreds of books and thousands of journal articles on sustainable business, business ethics, entrepreneurship, education, and management cover the topic of business and sustainability and its rapid evolution in detail; here we will introduce some of the common terminology and concepts only to show the breadth of approach and controversy that attends these difficult issues at the so-called "meso" level.

There are many economic and operational impediments businesses face that may inhibit pro-environmental behavior change. For instance, some economic analysts have pointed out the paradox that the less consumer demand for energy that is achieved by energy efficiency on the part of individuals and households, the greater the disincentive to adopting newer, greener technologies for energy generation within the power industry. This is because existing power plants using coal or petrochemicals have a long lifespan, and until they need to be replaced, if demand isn't increasing, companies are unlikely to build new power plants using these new technologies.[36]

Incentives for businesses for things like energy efficiency in their own operations can be influenced by the way institutions typically work as well as economic considerations. Let's take real estate and building construction as examples. When rewards follow the scenario known as "split incentives," decisions that improve energy efficiency can be stymied. Split incentives occur when one person is charged with accomplishing one goal, and another person a different goal, and the rewards of the two goals don't intersect. For instance, a residential landlord can pass energy costs on to tenants, but the landlord is not incentivized to spend the money to upgrade the energy efficiency of the rental units, because the tenants, not the landlord, get the savings on their utility bills. In many commercial construction projects, it's nobody's job to oversee and be rewarded for energy efficiency for the building as a whole; the plumber does one aspect, the drywall person another.[37] In many building projects, one set of people oversees capital costs, and another oversees operating costs. Operating costs will be less over the long term if there is more capital expenditure up front on energy efficient building techniques, but the use of these techniques increases the capital costs of the building. If the builder only cares about selling the building and therefore isn't concerned about subsequent operating costs, there is no incentive to spend the extra costs up front. Even in situations in which the company that builds the building will be operating it long-term, only if the participants interact during the planning phase, and there is financial reward to the company leadership that is passed down as some other kind of meaningful reward to the appropriate personnel, is an environmental goal like energy efficiency likely to be sufficiently motivating to change business-as-usual.[38]

Some behaviors arise mostly from informational failures. If you as a potential tenant knew that a building was energy efficient and that your heating bills would be lower, you might be willing to pay higher rent if you thought it would be offset by lower utility bills. But often this kind of information is lacking at the consumer level.[39]

Despite these structural and economic impediments to improving sustainability in real-world business practice, as awareness of environmental concerns has increased, companies have responded by finding ways to demonstrate a level of responsiveness to investors and to the public. In recent decades, the concept of Corporate Social Responsibility has become widespread, with the "triple bottom line" metrics of economic, environmental, and social performance indicators entering into the standard analysis of how a business is doing, typically reflected in its annual report.[40] Some companies have found that environmental goals can align well with corporate goals and can promote consumer loyalty and other business advantages.[41] From trucking to air transportation to tourism, entire journals have arisen to disseminate strategies and data regarding sustainability and its complex relationship to business priorities.[42]

It should not be surprising, however, that in many situations, economic, environmental, and social goals collide. One example is hotels that allow customers to "opt out" of daily room cleaning, with the idea this helps the environment, because of fewer chemicals and cleaning agents. This is good for the hotel financially, because they have lower expenses for housekeeping. But the housekeepers now have less certain work with unpredictable demand, and so may be hired per diem instead of full time with benefits. Additionally, cleaning a room that hasn't been cleaned in three days is more work for the same pay; the room is more likely to be a mess. Thus, the customer is making a choice for potentially pro-environmental motives, without being aware of the unfair social and economic consequences for the housekeepers—nor fully understanding the significant financial incentive for the hotel.

The effect of environmentally friendly practices on business profitability remains controversial, and, not surprisingly in light of the heterogeneity of businesses and markets, data show mixed results. On average, the introduction of sustainability practices into businesses may be a "break even" proposition, favoring some and being unfavorable to others, according to some comparative analyses.[43]

For this reason, some scholars make the case that market forces as they currently exist won't change the situation overall, at least in the critical decades in which greenhouse gas emissions need to be cut drastically. Nor, they say, will individuals in the wealthy classes do so by conspicuously cutting their consumption. In this view, the entire conceptualization of "nature" is seriously flawed. The root problem of progressive environmental destruction arises from capitalism, social inequity, land and resource "ownership," expansionist relegation of indigenous peoples to a powerless "naturalness," and exploitation based on self-serving social and economic paradigms.[44] Capitalist systems embody the right to own, control resources, exploit, and pollute. By this reasoning, only a true revolution—in which capitalism itself is replaced with a more just political system in which land and environmental resources are no longer considered part and parcel of capital to be exploited unequally—will appropriate environmental change happen.[45] Some of these writers criticize the idea that middle-class consumption is the problem, rather than the entire, exploitative capitalist system, and label "mainstream" environmental movements as "bourgeois environmentalism" based on comfortable, "middle class angst" about a degraded environment.[46] The conviction that improvements in quality of life must be coupled to ultimately unsustainable growth has been challenged by alternative economic models that advocate sustainable growth, or a so-called "steady state economy."[47] Others make the point that only through collective action, rather than individual change, can we have an impact that really matters.[48] Still others suggest that democracies will change more slowly than more autocratic societies, because democracies spend too long coming to consensus, politicians need to focus on getting elected in the short term, and there are too many special interests, especially those of economically powerful, enormous multinational businesses, in play.[49] When we talk about behavior change to overhaul social and economic systems, then, we are talking about people in leadership roles in these realms. But whose job is it to encourage these leaders to change? And what are their rewards for doing so? We will address some of what has been learned about this shortly.

Recall that when we are talking about organizational or political change, we are also talking about individuals. Whether we are talking about changing the behavior of people whose personal carbon output is multiple times the world average and need to reduce it to one tenth its current level to meet the IPCC targets, or those who are societal, governmental, policy, business, or social change leaders who may influence many others, *all of us are working with the same basic neural equipment*. We vary a great deal in our specific models of this equipment and how it has been shaped by experience, but nonetheless, there is no magic way around the fact that this particular threat is challenging for how we work as decision makers and even as change agents in all of our various spheres of influence.

Intersections: Micro, Meso, and Macro

Most research in environmentally relevant behavior, like most studies of behavior change in general, focus on what variables influence specific behaviors or attitudes in groups of individual subjects. The behaviors studied have focused most often on choices that individuals make in their domestic lives, rather than as influencers in their workplace or in a more collective or political sphere. However, the decisions made by individuals at the micro level are influenced by the landscape at the meso and macro levels as well, so these findings offer some useful information at the intersection of these different contexts.

In line with what we saw in Chapter 7, framing information so that the rewards of the behavior or risks of inaction are accentuated, and even using subliminal messaging, have been shown to influence pro-environmental behavior at the level of individuals.[50] Immediate feedback about consequences such as energy conservation and cost savings, using in-home meters or dashboard displays in cars, tends to be more effective than delayed feedback, consistent with how the brain instantaneously processes the reward value associated with a behavioral choice.[51] Similar to treating addictions, approaching several environmentally relevant behaviors at once and attempting to change behavior in a group context that includes social reinforcement tends to result in more lasting change.[52] We will explore specific examples of these strategies as we encounter them below.

As one example of what effectively motivates pro-environmental behavior change at several levels, we can look to one of the more intensively studied change opportunities for the individual consumer—home energy efficiency. While your brain may perceive this as an unexciting topic, we'll see if we can make it more interesting—*more rewarding*—to change this one aspect of behavior. The neural principles of how decisions are made and behaviors changed at the level of consumer choice also are applicable to decisions affecting many people, from our company executives to our legislators to our policy makers and media influencers, whose decisions in those spheres may have a wide scope.

Time for a New Kitchen?

Home energy efficiency provides a great example to study, because it touches on many real-world aspects of decision making and behavior change relevant to the environment, and involves both individual choice and organizational decisions. First, residential energy use is a significant part of the carbon problem in the high-income world, with a large amount of waste that can be decreased in a cost-effective way in many instances—so there is potential for reward. However, as rewards go, this is not a strongly positive one. Why? Cost *savings* are less rewarding than something gained that is immediate and tangible. Savings are theoretical and in the future, valued less than should be the case based on logic alone, but that's how our brain processes them.[53] Nonetheless, there is some reward to be had in this space, so it's got something going for it.

Second, there are multiple stakeholders—homeowners, energy companies, policy and government officials, environmental scientists and advocates, people selling products and services—so it's a good example of real-world complexity. Third, it's one of the better studied areas of environmentally relevant behavior change.

A lot of the research in how to get people to improve their home energy efficiency historically has been framed this way: if we could identify what makes people improve home efficiency (drivers) and what keeps them from doing it (barriers), we could get more people to do it.[54] But who are the "we" trying to get them to "do it"?

Somebody has to get a reward for incentivizing others to reduce energy consumption. Typically, there are local governmental agency workers who have to find ways to provide power to an increasing population. There are energy company executives who want more customers but don't want the expense of a new power plant eating into their profit margin, so want customers to use less energy per person. Finally, there are environmental policy makers, experts in the science of climate change who believe that mitigation is desperately needed to avert catastrophe and represents a just cause that is part of their professional role. These various people act as change motivators for citizens who may not think home energy efficiency is particularly important or exciting (read: rewarding) in the scheme of competing priorities. For these people, the first step is to successfully provide the motivation to make a change—that is, the "driver." It's also true that some private citizens independently may be motivated to change because of a sense of environmental concern; for such individuals, while the driver may be sufficient, there may be barriers to implementation.

Tellingly, there is a parallel here to what we learned in Chapter 3 from neuroeconomics—financial decisions are not always rational in the sense of prioritizing maximum monetary gain. Instead, many other factors are at work, reflecting how the brain is designed to make decisions based on multiple simultaneous factors, including attentional factors, social interaction, comparisons with others, fairness, empathy, and many other behavioral traits that arise from brain design. Similarly, much of the research on energy efficiency renovation has focused on financial incentives, as though this is the main way that people are motivated and make decisions. But that's a narrow and, as has been learned, incomplete way to consider behavior change strategies. The following example shows us why.

Governments, municipalities, and power companies in the United States, Canada, the United Kingdom, other European countries, and China have in recent decades provided a variety of financial incentives for home energy renovation.[55] But these perks have been significantly underutilized. Concluding that this was because of barriers, institutions have attempted, with some success, to make energy audits and renovations easier and more convenient.[56] The target behaviors might include installing better insulation and identifying and repairing air leaks, replacing older windows with double- or triple-paned versions, optimizing heating and hot water system efficiency, and changing lightbulbs to energy-efficient models. The company may encourage owners to arrange for an appointment for energy auditors to come to the customer's home, perform an efficiency assessment, provide prescreened contractors, and grant a discount on the work or a rebate on the homeowner's energy bill to help defray the expense.

Some of these incentives help. Having people come to you, and realizing savings, diminishes some of the barriers to even thinking about home energy efficiency—an attentional barrier. But when this approach has been studied, despite these financial incentives, it turns out that only 10 percent of renovations are performed to increase energy efficiency in households. In contrast, 90 percent of renovations are done to *make the house nicer for the owners in some way.*[57] They may want a new kitchen, or bathroom, or to add a new room. Because they're renovating anyway, adding some cost savings over time by increasing energy efficiency is a bonus, but is likely not the primary incentive. In fact, much of the mental processing and recognition about cost savings from energy efficiency happen after the fact, when the renovation is already completed.[58] So what's going on here?

"Home" Is Not a Rational Concept

Unless you're a house flipper or a real estate investor, for most people, your home has a lot of emotional meaning. There is an entire branch of environmental psychology that deals with humans' sense of place as a strong tendency to identify with a location in which you live or were raised.[59] For many people, home improvement, including renovation, isn't an event—it's a process. Many people spend enormous amounts of mental energy, time, and resources to continuously make their homes pleasing; they are effortlessly motivated to do so and find it satisfying. This tendency to decorate and redecorate in order to both optimize your enjoyment and portray yourself to the world is an entrenched human trait. Clothing and your home are ways you present yourself to other people, and these take on a good deal of importance to many individuals.

But energy efficiency? Yes, people may know it's important, and feel good if they do it; but they are much more likely to do it in the setting of a renovation done for another reason—in the words of behavioral science, an entirely different goal-directed behavior. The kitchen would be nicer with new cabinets and an island. The living room entry could be moved and it would free up more room for glass doors to the patio. This basement could be a workshop. This attic could be a studio and a spare guest room for company.

Now, this is fun—this is *rewarding*. Positive is better than negative. Planning and visualizing and dreaming of things that are pleasant upgrades provides a spritz of dopamine even during mental anticipation, like we saw using neurochemical probes in Chapter 2. Getting new "stuff" is a built-in big reward; looking through home improvement and decorating magazines for ideas—this is not work. And if energy efficiency can hitch a ride on this motivation, with some cost savings (now I can afford that light fixture I wanted!) and good feelings from being good citizens as a bonus, all the better. If the house is warmer, less drafty, and if the new windows look great and don't need to be painted every year—even better. And, if I learn about energy efficiency strategies and which windows do the best job from my next door neighbor, rather than some utility company guy or sales person, even better still—I am, on average, four times more likely to incorporate the energy-efficiency features than I would have been otherwise.[60] Information from people I know has a bigger effect than information from "experts" who are strangers—just as in the case of Ebola in West Africa that we encountered in Chapter 7. Social affirmation from others like me is a powerful motivator—even if it's to make my windows look as nice, or maybe even better, than those of that couple down the street of whom I'm a little jealous—they're nice, but a little too perfect, if you know what I mean. This I understand, and it doesn't require a carbon calculator.

With this example of improving home energy efficiency, we begin to see the complexity of changing behavior relevant to the environment. We might see parallels to our homes in how an executive or director personally identifies with a business or institution; how it "looks" and acts and its reputation to outside observers may provide potent motivations in decision making, beyond purely financial considerations. As deeper understanding of climate change and its relation to human behavior has evolved, researchers have had to adjust their focus and use new approaches to learn what influences people and what might work to best motivate change. Let's next look at some of the ways they have studied these questions, and what we can learn from their findings.

Predictors of Pro-Environmental Behavior

When one is trying to get people to wash their hands before eating to prevent infectious diseases, or exercise more, or use sunscreen, or stop using addictive substances, the behavior is specific, and the outcome directly affects the person whose habits are being changed. In behaviors that affect the environment, there are many behaviors a person can change, all of which require that person to change customary behavior for goals that affect the individual far less directly and less immediately. This makes environmental behavior both harder to change and harder to study.

One way that researchers try to simplify complex behavior is to build predictive models. Market researchers do this for products; if a customer is in the market for a new car, what profile of people tend to go for which of your models, compared to the competition? With this information, you may be able to target your marketing to people who might be most inclined to consider your brand. Likewise, when trying to encourage specific types of pro-environmental behavior, researchers have gathered data that predict what kinds of people might be most inclined to be receptive to change, and which specific kinds of change might resonate for them. For instance, people's sense of altruism, valuation of non-human species, sense of personal responsibility and agency, religious views of the relationship of people to nature, and concept of the interdependence of human-environment relations all may help predict which types of pro-environmental behavior a particular individual may be most likely to undertake. Those with a strong worldview of the importance of other species may respond to issues threatening biodiversity, while those concerned about other people might be motivated to change by climate justice issues. Conversely, people whose values focus on self-enhancement, obedience, self-discipline, and family security are less likely to prioritize environmental issues.[61]

There remains a gap between pro-environmental *intent* and the actual impact of specific behavioral *actions,* even among people who have pro-environmental attitudes.[62] However, in this context, environmental knowledge does seem to help. Large surveys involving tens of thousands of people have corroborated that in general, people who are better off financially, better educated, in mid-adulthood, female (slightly more than male), more knowledgeable about environmental issues, and have a social / altruistic perspective are more likely to report that they currently engage in high levels of pro-environmental behavior, such as changing their transportation, decreasing household energy and water utilization, reducing solid waste and recycling, eating local food, purchasing energy from renewable sources, and reducing air travel.[63] Personal tangible experience also plays a role. People in flood or drought zones or those who have experienced weather events directly have a higher rate of belief in the role of climate change in alteration of the natural world compared to those without such experience, and they express greater support for direct action to mitigate threats.[64]

But what people say they'll do and what they actually do are not always the same. In fact, in many studies, not by a long shot. It is easier and cheaper to use questionnaires to survey what affects people's attitudes than it is to do field work to document whether their real-life behavior actually changes under given influences. For this reason, even the best predictive models only have a moderate relationship to actual, measurable behavior change. In part, this mismatch results from the fact that people can know about climate change intellectually, but they have many justifications that can be utilized when it's just easier to keep doing what they're doing without making bothersome changes or feeling awful.[65] As we've seen, the brain is equipped to adapt based on what's happening right now, and can find ways to attend selectively.

Other reasons for the disparity between intention and action arise from structural barriers; you might want to take public transportation, for instance, but it just may not be available or practical in your daily life.[66] Thus, decreasing barriers to make behaviors easier—like putting a recycling bin in every office, not down the hall—works better than asking people to do things that are more work or require more time. And as we discovered in Chapter 6, even well-intentioned people often just don't have the knowledge about which behaviors actually have the most environmental impact, so they often put their efforts into activities that don't make as much difference as others might.[67]

Successes—and Some Failures

With our understanding of the complexity of these issues, we now can examine several specific strategies to encourage pro-environmental behavior and can begin to see in context what researchers have learned about how well they work.

Environmental Communication

While we have noted that information alone doesn't correlate well with changing behavior, it is felt by many researchers to be a precondition, or primer, of behavioral change. This has given rise to an entire field of *environmental communication* that investigates what is being used and what works best to spread awareness of the problem as a first step.[68] Evidence suggests that messages are most effective at engaging their audience when they are concrete rather than abstract, and have features that allow recipients to identify with the situation, overcoming some of the "discounting" that "this will happen in the future, or far away."

We have focused on change occurring in the brains of individuals, but of course, many individuals can be influenced by the same information if it is broadly disseminated. Books have played a major role in shaping the opinions of the public, and the most effective share some of the characteristics noted above. Rachel Carson's *Silent Spring* made the intertwined web of pesticide toxicity tangible and its unchecked consequences both frightening and sad—but offered a hopeful way forward. Bill McKibben's *The End of Nature,* the first general audience book on global warming, opened the eyes of people outside the environmental science world to the threats and challenges that have only become increasingly apparent since that alarm was first sounded.[69] In its subsequent reissues and in additional books, McKibben has documented the widening disconnection between the acceleration and obviousness of the problem, and the denial-laden response of the public and of their leaders, thus spawning activism beyond just journalism.[70]

Illumination of the hidden motives and behind-the-scenes manipulations practiced by industries linked to climate change occurred even before their environmental effects were recognized. Ida Tarbell's history of the Standard Oil Company showed the unfair practices that consolidated power in the hands of the few at the expense of many, as well as the cognitive dissonance of its leaders, described in vivid prose to which her audience could relate both intellectually and emotionally:

> Those theories which the body of oil men held as vital and fundamental Mr. Rockefeller and his associates either did not comprehend or were deaf to. This lack of comprehension by many men of what seems to other men to be the most obvious principles of justice is

> not rare. Many men who are widely known as good, share it. Mr. Rockefeller was "good." There was no more faithful Baptist in Cleveland than he. Every enterprise of that church he had supported liberally from his youth. He gave to its poor. He visited its sick. He wept with its suffering. Moreover, he gave unostentatiously to many outside charities of whose worthiness he was satisfied. He was simple and frugal in his habits. He never went to the theatre, never drank wine. He gave much time to the training of his children, seeking to develop in them his own habits of economy and of charity. Yet he was willing to strain every nerve to obtain for himself special and unjust privileges from the railroads which were bound to ruin every man in the oil business not sharing them with him.[71]

Tarbell thus gained recognition as one of a cadre of "muckrakers" who illuminated corporate corruption and greed in a variety of industries. These widely read works helped shift public attitudes, spur protests and social movements, and influence legislation, including anti-trust laws and food safety regulations.[72] In the same vein, in this century, books like *Merchants of Doubt* exposed the widespread public disinformation campaigns waged by the tobacco and fossil fuel industries based on deliberate manipulation of scientific data and deceptive consumer marketing.[73] Such exposés become part of what influences individual people as they make decisions relative to those companies, based on their public image and credibility sometimes as much as the downstream effects of their products. As in the early twentieth century, information that reaches many people through books, films, and other media can turn public opinion even against entities holding considerable power and can catalyze collective action.

In journalism and broadcasts about environmental issues, news that shows actual people who are victims of weather events attributed to a changing climate tend to garner a more robust response than do abstractions like images of political figures, and narrowing the "psychological distance" more effectively changes people's attitudes toward a problem.[74] Images can be pivotal; the "Earth Rise" photograph from the 1968 Apollo 8 space mission has been credited with enhancing widespread recognition of both the beauty and the fragility of our shared and finite planet.[75] Pictures tend to work better than words at garnering an emotional response—think of the polar bear trapped on the ice floe. The images that scientists and experts think are most compelling, sometimes showing facts and graphs, are not necessarily the ones that elicit a response from the general public.[76] Functional MRI studies show that people tend to have empathetic responses to photographs of degraded environments similar to but generally less intense than those for animals in distress.[77] Under experimental conditions, the power of such emotional entreaties to change behavior is enhanced when paired with an immediately available relevant actionable choice, such as purchasing a low-energy light bulb.[78]

How scientific ideas and the adoption of pro-environmental behaviors spread throughout larger groups of people also has been studied. There are three stages: knowledge, persuasion, and adoption. Knowledge of new things tends to spread by mass media, but persuasion about them tends to result from the input of people known by the listener. Adoption tends to occur if the new innovation or behavior has an advantage to the adopter compared to alternative behaviors.[79] Thus, beyond just knowing about something, actual behavior changes relevant to the environment have to meet this bar—changing must have a perceived advantage (reward!) to the individual in order to be adopted and "spread" through society.

Opportunities for environmental education and behavior change are provided to a wide audience by electronic media, and effective climate education can utilize some of the same strategies used in marketing that we saw in Chapter 5.[80] Online information can be used as normative anchoring for people, especially when viewers see that many people affirm a point of view.[81] Numerous websites have arisen to help people with all kinds of information—carbon calculators, product analyses, company comparisons, many with an environmental focus, some also with a social justice component. But it's up to you to go look for it. If you want to buy a new cleaning product, a new outfit, or a new couch, you are likely to be able to find something to inform your choice on environmental grounds.

On the other hand, mass media has played an outsized role in perpetuation of climate change denial and skepticism. The roots of this movement have come from the petrochemical industry, conservative funders, other economic or political interests, and a limited number of scientists with views outside the majority in their relevant fields. These views have been propagated by bloggers and "free speech" advocates, thus sowing doubt on scientific consensus that is otherwise increasingly consistent.[82] Attempts at journalistic balance by including the views of scientific contrarians has been judged by some commentators to have backfired, playing a role in skewing the public's perception and undermining recognition of the degree to which a consensus based on overwhelming factual scientific evidence already has been reached about climate change.[83] As in other science-based contexts in which findings bumped up against economic or cultural self-interest, the question of climate change has served as a platform for some groups to undermine the tenets of the scientific method itself. These strategies include deliberately misinterpreting the constant self-questioning and built-in iterative processes inherent in scientific inquiry that are designed to continuously refine and uncover greater depths of verifiable factual truths. Such commentators capitalize on the expected uncertainties and controversies about the details within emerging scientific findings in order to sow doubt about the overall big picture consensus. Science isn't perfect, nor are scientists, but as a system for finding new knowledge, it's generally believed by most observers to be the best process available, and it continues to pass the test of time.

Attempts to create appropriate evidence-based narratives that the public can understand have been actively explored by those in the climate change arena to try to bridge the gap between environmental knowledge and environmental action.[84] Thus, scientific research on effective environmental communication and behavior change confirms that just telling people "the facts" is an oversimplified and ineffective approach. Here too, how information is framed—presenting a behavioral choice as leading to something rewarding, or avoiding some frightening risk—changes how people respond to it; this can make the information more compelling and more likely to lead to a change in behavior. For instance, framing climate change as a public health issue that causes harm to health and well-being can motivate concern and public engagement, especially toward a positive goal to achieve a more vibrant and healthier planet.[85] Showing a business leader that there are significant cost savings to be had with processes that reduce waste and the associated cost from hauling it away frames the environmentally preferable behavior as something positive for the bottom line—without even needing to convince the leadership that climate change is a real problem.[86] A hotel can frame that reusing your towel rather than replacing them daily is good for the planet and that 75 percent of people who stayed in this room reused their towel, providing a competition gauntlet. The hotel's motivation? Something positive to say in their Corporate Social Responsibility annual report, good public relations, and perhaps most importantly, the hotel saves money, because they don't have to wash and replace the towels every day. Aligning goals this way may be the most effective route to behavior change. (Whether this is completely ethical or not is another question raised by some authors—because, among other issues, what if what the customer is told is not true?)[87]

Other analysts have emphasized that to help people understand the necessary behavioral changes on everyone's part to address climate change, the solution needs to be framed not as a physical process, but a social one, in which people's choices, behaviors, and interactions are key.[88] Conceptual and rhetorical changes, such as the Gaia Hypothesis, Whole Earth, and Deep Ecology paradigms are designed to try to express the immutable interrelation of humans and the rest of the natural world; persuading the general public along these lines is a well-described challenge.[89]

Additionally, information and images that also provide guidance about practical, meaningful action are important.[90] As we saw in Chapter 3, efficacy is rewarding to humans. Things that just make you feel bad about a problem but don't give you a realistic way to act that you believe can make a difference don't effectively change behavior. They may even backfire, causing people to just give up. Focusing on climate change action as a moral imperative and emphasizing efficacy and hope can help reduce "compassion fade" in the environmental realm.[91] Similarly, interventions that encourage people to believe that they have the ability to control their behavior in decisions with environmental impact are more likely to result in pro-environmental behavior.[92]

Social Rewards

Capitalizing on what's been learned from general strategies for behavior change, some programs have patterned pro-environmental behavior change efforts on those used for other difficult habits. As we've seen, in this "wicked problem," multiple behaviors have to change, the behavior (not just the intent) has to actually occur, and the behavioral change has to be long-lasting—basically, substituting new habits for old ones. Positive works better than negative—people who behave environmentally and feel pride, not guilt, are more likely to sustain a behavioral change.[93] An example of one such multidimensional "small-group" program is the international Eco-Team Programme, based on the core behavioral change strategies of information, feedback, and social support to tackle barriers to pro-environmental behavior at the level of individuals in households. More than 20,000 households have participated.

The program organizes neighborhood teams led by a local block leader. The teams meet regularly for about eight months and elicit an expressed commitment by participants—one of the behavioral strategies (public commitment to a behavioral change) used successfully in other contexts, like Alcoholics Anonymous. Six pro-environmental behaviors are the targets, chosen by the specific team from a list of about 100 options. At the meetings, information is shared, and ideas are generated by the group of friends and neighbors. Progress is tracked on relevant environmental metrics, such as the weight of trash generated and the amount of energy and water consumed by each household. Results are compared to other teams in the area, so that feedback is promptly available, and there is an element of competition. The program takes place in a supportive social environment with encouragement by others on the team, and it includes some team socializing.

So how effective is this approach? Compared to behavior change of people who just attended informational lectures, the EcoTeams approach has led to lasting change in pro-environmental behaviors two years later, with significantly more change that was seen in an information-only control group.[94] In fact, behavioral changes seemed to increase with time and expand beyond some of the initial targeted behaviors, both in the Netherlands study and that in the United Kingdom, as well as in other similar small-group interventions.[95] The question of whether "response generalization" occurs from one behavior to a related one has been raised in pro-environmental behavior change studies, and often the results are negative when a single behavior change was targeted. But in this specific multidimensional approach, there is some indication that crossover occurs. Pro-environmental habits appeared to replace old habits, and identities as environmental advocates seem to have become solidified, even when compared to people with similar environmental attitudes at the beginning of the study period.[96]

Nonetheless, to be maintained, behavior has to be rewarding, and it must be more rewarding than the alternative. Unless the social reward is sustained, the behavior may fade; some researchers have noted "rebound effects"—people do something pro-environmental, and then they feel justified doing something more environmentally damaging, as though they've built up "credit" on their footprint—like driving their hybrid car longer distances, or keeping their thermostat higher now that they have a heat pump, cancelling out the gains.[97] But when people make a commitment to change that includes the social reward of public affirmation and shared goals with others, it does tend to stick—if the alternative behavior is rewarding, and even more so if it can be made to be fun in some way. If kids are recycling, say some, let them smash the cans. If you are commuting on the train rather than driving, find a way to use the time that is enjoyable or productive and becomes part of your new routine—a new habit.[98] *If it's not rewarding, we simply won't do it.*

Other efforts that fall under the category of social rewards include those from grassroots and advocacy organizations. These groups often combine environmental advocacy with other "drivers." Many organizations, such as the Sierra Club, the Nature Conservancy, National Wildlife Federation, and the Audubon Society, share motivations to protect the environment because of their commitment to natural spaces and wildlife. Grassroots groups with many local chapters provide members and donors with tangible collective tasks and the increased efficacy and support that group participation can provide. Similar approaches have been used at the town level.[99] Organizations like the Appalachian Mountain Club, Save the Bay, and local land trusts and waterways organizations focus on specific geographies or habitats. Ducks Unlimited protects habitat by appealing to hunters—a novel reward / environment partnership.

Concern for the future of their children—arguably one of the most potent forces known—motivates those involved in organizations like Mothers Out Front. This group enjoins community-based chapters to identify and tackle local environmental projects, like stopping the underground gas line leaks that are both dangerous and energy-wasting all around the greater Boston area. Members and chapters provide one another with mutual support and share strategies. Additionally, local groups collaborate to identify projects and policy initiatives at the state level, and state groups do the same to identify national issues, adding collective action and political change to their individual change agendas. The rewards include relationships, community, a renewed sense of agency, and the opportunity and mentorship to develop leadership and political skills that can be used to influence more long-range systemic progress. Thus, this approach joins a visceral and lasting drive—concern for one's children—with an immediate problem to be solved with others, with skill-building and social support for action at multiple levels. Such a strategy brilliantly checks many of the boxes we've encountered for potentially effective behavioral change that also can spread socially and politically over time.

Groups like the Environmental Defense Fund and Union of Concerned Scientists engage in research and advocacy based on scientific data, and they provide both practical and advocacy support for members engaged in environmental action. Many organizations have educational and political activities, with some being quite powerful as lobbyists and sometimes plaintiffs in legal proceedings. These and hundreds of other organizations provide ways for people to connect to others and validate their shared commitment, sense of identity, and purpose toward "doing the right thing," which can be a powerful social motivator and effector of change.[100]

Carbon and Health Initiatives

Carbon taxes and carbon trading are macroeconomic measures designed to reduce greenhouse gas emissions. But some experimental programs have used this strategy at meso and micro levels. The argument for carbon taxes is that the costs of carbon emissions are not just what the petrochemicals and other greenhouse gas sources cost in market terms. Instead, calculating the cost also should include the more long-range costs of the accelerating effects of climate change on all of society. Thus, these more long-range, life cycle costs should be factored into the use of carbon-emitting processes and energy sources by businesses and industry, rather than having the industry benefit from cheap petrochemicals while taxpayers pick up the environmental costs later. Similarly, individuals who choose to engage in high-carbon lifestyles would pay for some of the more long-range costs of this choice. The revenue from the carbon tax could support things like development and implementation of alternative energy sources, improved technologies, land restoration, waste reduction, aid for climate refugees, weather-related disaster expenses, food and water redistribution, and other environmental mitigation (prevention) and adaptation (response) efforts. These approaches have many strong advocates and have been implemented—and occasionally reversed—in many countries and provinces, while they have met resistance in the United States and China. Not surprisingly, they represent a major change from the status quo that generates considerable controversy at the levels of policy researchers, industry representatives, politicians, and individuals.[101]

In simple terms, when applied at the industry level, carbon taxes are fees paid by companies for every ton of carbon emissions they produce. A cap-and-trade system provides "allowances" that companies can buy and sell for each ton of greenhouse gas they emit. These types of measures reduce emissions by aligning the economic incentives of the company with the environmental incentives. They also encourage entrepreneurial development of low-carbon technologies. Critics argue that the undue burden on businesses will cripple economic growth and that taxes collected for carbon may be used by governments for non-environmental purposes, thus potentially diluting their purpose.

In contrast, *personal* carbon taxes are applied at the individual consumer level, usually to fuels or energy charges, or sometimes to products like automobiles or appliances that utilize energy. This is a "short-term pain for long-term gain" scenario, and critics from various points of view argue that this approach penalizes consumers for things that are "not their fault," doesn't benefit them directly, or (from those who don't believe in climate change) is aimed at a problem that doesn't actually exist. Some argue that consumers don't have the information to understand which of their own activities are associated with what degree of carbon emissions, so that personal carbon taxes applied to individual purchases (for example, added as a tax to gasoline at the pump) don't always help consumers recognize and decrease high-consumption activities.[102] However, advocates point out that those who use more, pay more, and this kind of tax is a direct approach to reducing household footprints.

Some researchers have suggested that to really push individual behavior toward pro-environmental choices and toward personal health and well-being and financial security, all of these things should be linked, in a convenient, rewards-based method. Two such experimental programs are the Norfolk Island Carbon Health Evaluation (NICHE) project in Australia, and the Carbon, Health, and Social Saving System in British Columbia, Canada.[103] Such programs use elements shown to be successful in other contexts of difficult behavioral change, including direct links to personal health and well-being, economic incentives, social norming and reward, reinforcement of efficacy, and decreased barriers by the use of conveniences like special credit cards for carbon that increase consumer knowledge and transaction transparency. How successfully such programs would scale to a larger and more heterogeneously motivated population is at present an unanswered question.

Environmental Behavior without the Environment

There are some approaches that attempt to get people to change behavior not because of knowledge about or commitment to environmental causes, but because of some other personal benefit. Such approaches bypass the "information" part of the picture altogether. This has been described as the "honeybee" approach—bring people along by attracting them to a behavior that includes a self-serving motive. Need a new building for your business? Using a more energy-efficient design will save you operational costs.[104] Want the health benefits of exercise but don't have time? Biking to work may fit the bill.[105] Redecorating the guest room? Shopping for quirky thrift-store furniture saves money, provides the fun of the hunt, and shows your style credentials; the fact that reusing things is good for the environment is a secondary perk, if recognized at all. This general approach takes advantage of what is known to be rewarding to people based on evolutionary and behavioral principles, and in essence it "markets" or "frames" behaviors to encourage people to make a choice that is also, covertly, environmentally advantageous.[106]

It's worth pointing out that the same behavior can be categorized differently, depending on the specific motivation. For instance, spending money to insulate your home, if you're financially strapped and can only pay for it by doing without something else you need, is a social dilemma problem—it's for the collective good but is not what you personally need most right now. Alternatively, if you choose to prioritize insulation because you care about the environment and know it will lower your energy consumption, and it's an investment you can afford—then you're making an environmental choice. But if you're doing it simply because you want a warmer, more comfortable house and don't care about global warming, then the behavior falls into the honeybee category.[107]

Like many other strategies to promote behavior change relevant to the environment, suggestions about playing to self-serving advantages (better health, lower cost, less effort, nicer living space) don't clearly delineate *who* should be the instigator of such behavior change and in what kind of systematic program or policy. More often in policy or consumer choices, the environmental goals are not covert; they are one of a number of touted benefits of a behavior-change effort, even if not at the top of the list—such as financial incentives for home energy efficiency discussed in our example above. We will examine other examples that use the honeybee approach in Chapter 9.

Nudging and Budging

Behavioral nudges that present choices in a way that favors the options that have socially desirable consequences (see Chapter 7) also have been applied extensively to environmental behavioral choices. Besides being the focus of considerable academic research, governments as well as industry have utilized this approach. In a study in Ireland, electric customers could use electricity whenever they wanted but paid more for peak usage, and they could get immediate feedback about their energy use from "smart meters" in their homes. Besides reducing peak usage, customers also reduced their *overall* energy consumption.[108] Use of green energy defaults—obtaining energy or electricity with a higher proportion of renewable sources, with more work being required to change to an alternative provider—has been shown to be a very effective strategy to change behavior, even when clients have different political leanings or have to pay more—just because it's easier to choose the default.[109]

Social norming and competition ("others in your neighborhood used less energy than you did") and product energy efficiency labeling with an appropriate "anchor" to encourage the most environmentally friendly choice have shown some efficacy in improving household environmental behavior.[110] Researchers who compared different framings of informational messages on subjects' willingness to decrease red meat consumption in Canada found that those who received a message framed as a social norm ("people are making dietary changes to reflect their feelings about these impacts") were most willing to consider changing their behavior.[111]

In the agricultural sector, a study of practices affecting water pollution in Scotland showed that farmers were more likely to change their behavior with a voluntary nudge (information sessions, farm visits by coaches, voluntary suggestions, help with implementation) than a regulatory budge (regulations requiring specific practices). This behavior was thought to reflect a shared sense of ownership of the pollution problem rather than engendering resistance to additional regulations, which some farmers feel already exist at a burdensome level.[112] Similar motivations have been found in direct interviews with farmers about their willingness to enact environmental practices.[113] As we have seen in many different contexts, agency is rewarding.

But some analysts believe that regulation, in the form of budges rather than nudges, will be needed to change behavior relevant to environmental concerns. Nudging alone may not produce a large enough impact, and inherent biases against change in an unknown context with unclear benefits are difficult to overcome.[114] Current economic forces seem unlikely to support changing to pro-environmental behavior in isolation; regulatory interventions will be necessary to make the greener choice easier, cheaper, or competitively accessible.[115] Additionally, objections to nudging in the context of public health also apply to the environmental context, that is, that nudges covertly manipulate choice and thereby raise ethical concerns.[116] Exactly how psychological understanding of nudging gets translated into policy remains incompletely worked out, but it can involve framing, persuasion, and norm creation. To be successful, all of these require some trust between the public and the government agency involved, and norm creation in particular occurs by people interacting with one another, which governments can facilitate.[117] Taken together, these various analyses suggest that nudging and regulation both are necessary tools in this complex grand challenge.

Can Nature Itself Change Our Behavior?

Take a moment to think about your favorite place.

Across cultures, when people are asked to identify their favorite place, something they think is worth preserving, many people cite a place that is familiar. It is most often a family home or area nearby, and usually the favorite place includes some access to nature. Whether in Ireland, Senegal, or the United States, when asked why a particular place is their favorite, most people respond that it is a refuge, brings them comfort, and allows them to recharge.[118] Familiar natural environments are perceived as physical spaces that are worth protecting and saving by people who have experienced and become connected to them.

If the biophilia hypothesis we explored in Chapter 4 is correct, that humans are in fact inherently drawn to and rewarded by interaction with the natural world, it would seem obvious that one of the *rewards* of pro-environmental behavior is preserving the opportunity to enjoy nature itself. Conversely, there is a body of research suggesting that exposure to nature is associated with engaging in pro-environmental behavior. Is there evidence that simply exposing people to nature is a meaningful strategy to increase environmental behavior, and so, is the effect to a degree that might matter in averting climate change? Is the reward of nature enough to change behavior?

There is considerable evidence that exposure to nature correlates with pro-environmental attitudes, and some evidence that exposure to nature correlates with pro-environmental behavior, but detailed cause and effect as well as the necessary "dosage" remain incompletely understood.[119] Let's consider how some of these studies are done. Subjects often are university students (for instance, the study noted above on favorite places surveyed students from three different continents), and sometimes broader groups of people, usually recruited by online advertisements or government surveys. Questionnaires on pro-environmental attitudes might ask to what extent subjects agree or disagree with statements such as "I value nature and think it should be protected" or "I believe climate change is a major problem." Survey questions on pro-environmental behaviors might ask how often participants engage in recycling, take part in outdoor activities, or donate to conservation organizations. The astute reader will notice that studies differ in the extent to which they include the high impact behaviors we discussed in Chapter 6: "What model of car do you drive? How many miles do you commute, and how many of those are on public transportation? How many plane trips did you take last year? How many times per week do you eat meat? How many children do you have?" Additionally, because many people who agree to complete such surveys want to be good environmental citizens, it's hard to know how close their survey answers are to the reality of how they live; it's much harder to measure actual behavior.

Let's also touch on another issue relevant to interpreting this kind of research—the very definition of "nature." How this word is conceptualized differs among individuals and locales. In the United States, Canada, and Australia, for instance, many people associate the idea of "nature" with "wilderness," which is closely bound up with national history and cultural identity. "Nature" in Europe, India, and the Arctic may be perceived very differently and conjure different connotations based on local culture and modes of interaction.[120] Does a kitchen garden count as "nature"? How about a city sidewalk on the edge of a park? What about working in agricultural fields—is that "exposure to nature"? Driving on highways where there are mountains in the distance? Living in a building covered with vines? It's a fuzzy term, reflecting a contextual concept. Some theorists have argued that conceptualizing "nature" in a generic and geographically undefined way ignores the strong sense of emotional attachment to specific places that people tend to develop from experience, a sense of place. "Connectedness to nature" may be more accurately described as connection to specific places that contain some natural elements. Therefore, it may be oversimplified to think that just exposing people to nature in some general way will change behavior effectively.[121]

Research on the effect of nature on pro-environmental behavior is further complicated by social justice considerations. Widely differing priorities centered on land use and ownership can conflict among governments, landowners, land users, businesses, and environmental advocates. If the "nature" someone else is trying to protect is where people with less economic or political power live or make a living, its "protection" may pose a threat to a sense of place, means of income, and a way of life. For these reasons, some authors have argued that both scientific and cultural conceptualizations of nature and environmentalism are necessary to bridge divides, and to enable fair and just progress in the critical human / nature interaction changes that are needed to avert accelerating environmental crises.[122]

Exposure to Nature

With those caveats in mind, let's look at one reasonably well-studied topic, exposure to nature during childhood. Does it influence pro-environmental behavior? This can be examined by questioning children about their experiences in nature and correlating these with their environmental attitudes and self-reported behaviors. Another approach is to look at groups of youngsters exposed to specific activities, like camp, environmental education programs, or wilderness experiences, and asking similar questions, sometimes before and after the intervention. Some studies ask teachers to observe and measure specific behaviors, like recycling. Few studies, though, follow children long-term and determine which behaviors persist into adulthood. To get around this limitation, other studies look backward—they assess environmental attitudes and behaviors in adults and ask them about experiences in childhood, trying to get at which factors might have been formative for those who make pro-environmental choices in their adult lives.[123]

There is evidence that early life experience in "wild" nature—playing in woods, hiking, fishing—influences environmental attitudes and "human-nature connectedness," as well as the frequency of self-reported pro-environmental behaviors.[124] However, specific studies have surveyed participants about a varying list of pro-environmental behaviors, such as making voting decisions based on a political candidate's environmental views, participating in activities such as Earth Day clean-ups, preferring to spend time outdoors, buying local / organic food, recycling / composting, and taking public transportation or biking. Thus, how people respond may be influenced by other practical factors such as availability of specific activities. As a different metric, adults who report that they spent time in natural spaces during childhood tend to spend time in similar spaces during adulthood, controlling for other variables like socioeconomic status, age, gender, and locale.[125] Of course, all these findings represent a correlation, not a cause. Rather than the exposure to nature during childhood causing pro-environmental attitudes and behavior, children who spend a lot of time in nature might just have started out predisposed to liking nature more than other people, and that attitude persists into adulthood.

Despite the variability in methodology, there are some consistent themes arising both from research supporting the biophilia hypothesis we explored in Chapter 4 (that people are inherently predisposed to an affinity with living things) and research investigating whether exposure to nature increases pro-environmental attitudes and behavior. One common theme is that it is unstructured free play in wild, unstructured natural spaces, alone or with others, that seems to correlate with the highest valuation of the natural world. The setting does not have to be spectacular. A field or patch of woods or a creek near home is what many people remember, with frequent, often daily, ability to roam and explore there.[126] (Recall biologist E. O. Wilson spending hours in the swamps looking at insects). Having parents who have a high regard for nature, often sharing experiences in natural environments with family members, also predicts pro-environmental attitudes and behaviors.[127] Of course, it's also possible that some parents and children share a genetic predisposition to appreciate nature. Whatever the combination of heredity and learning, together these observations about the effect of free play and exploration, as well as parent's or friend's support for such activities, align with the neurobiological tendencies discussed previously: that agency is rewarding; that children are predisposed to learn from their elders; that social bonding is powerful; and that nature provides its own panoply of rewards, including effortless attention, restoration, respite, and adventure.

But there are some important caveats to these findings. First, not all—and in some studies, not even most—kids say they have a favorable impression of nature, even when they're exposed to it.[128] In particular, some urban kids find it frightening. This doesn't mean they won't look back on childhood nature exposure and value it later in life, but that's not been studied prospectively.

Second, context matters. Children who grow up working in agricultural pursuits to help their families in rural environments don't on average have the same love of natural spaces for recreation that kids who exclusively *play* in nature do.[129] A study of adults in Switzerland who worked in the forestry industry found similarly lower restorative benefits from nature exposure, attributed to not getting the benefit of "feeling away" from everyday life that an excursion into natural surroundings does for people who aren't working in it.[130] These observations dovetail with the criticism of mainstream environmentalism we encountered in our discussion of social justice—that "preservation of nature" by the middle class may be motivated by their desire to protect it as a commodity for their own enjoyment without cognizance that some people are working hard to make a living from the land, and may view the whole question quite differently. Then again, while the countryside may not be as attractive as a getaway for people who live there, on average, rural residents are more likely to resist changes to their surrounding landscape compared to people in the city. Urban dwellers are somewhat more open to the possibility of rural landscape changes occurring in order to meet the changing needs of the population, such as for renewable energy from wind farms.[131] As we saw in Chapter 3, we are designed to be rewarded for both novelty and for familiarity, and if a place is familiar, a strong place attachment may strengthen the motivation to preserve it unaltered.

While exposure to nature within urban settings has been linked to many benefits, as noted in Chapter 4, does it also change people's behavior in pro-environmental ways? This is challenging to study, but a few investigators have contributed some observations. Interviews of subjects in England showed that only visits to the countryside, not those to urban green space, predicted pro-environmental behavior.[132] Exposure of urban children to birdwatching and involvement in school gardens are described by many students and teachers as promoting cooperation, calm, connection to nature, and a setting where children with learning or social differences can thrive on a more level playing field with other students, but linkages to long-term environmentalism are unknown.[133] Similarly, long-term effects of school play spaces or natural urban landscapes on pro-environmental attitudes and behavior remain unclear, although there is some evidence to suggest its benefit in behavioral and cognitive spheres, and in the appreciation of nature generally.[134]

Overall, then, several lines of evidence suggest that children may learn best how to be environmental activists by learning to value nature through direct positive experience in it, often through free play and exploration; by acquiring a sense of agency through participating in successful pro-environmental actions; and by the experience of making a commitment to nature preservation. This is often embodied by a connection to a specific place, group, or cause. These principles follow those for social learning and ecological psychology that are applicable to both children and adults.[135] A study by Bixler and colleagues surveyed nearly 2,000 American students in public middle and high schools on their childhood activities, competence and fear of nature, environmentalism, and anticipated occupations. Wilderness experience correlated with comfort in and regard for nature, but, interestingly, not necessarily to an intellectual or cognitive interest in environmentalism.[136]

However, it's clear that nearly all studies along these lines show correlation, not causation. Because someone did something in childhood and values that activity in adulthood might mean the experience influenced the later attitude or behavior, or it might mean that people who liked something as kids keep liking it when they become adults. This distinction is relevant, because much of the supposition behind these kinds of studies is that since adults who experienced nature as children value it, then exposure to nature is a good strategy to turn children into environmentally concerned adults who are likely to undertake pro-environmental behaviors. However, you may not be able to produce environmentally oriented adults by exposing them to nature if it isn't in them from the start. The only way to really know would be to take a large group of a variety of types of children, randomly assign half of them to frequent wild nature experience and half to something else as a control group, and then follow them for a decade or two to see what happens. Studies have been done in the short term—what happens to children after camp, for instance—but long-term studies are much more difficult (and expensive) to do.

To approach the question differently, some "looking backward" investigations have focused on active environmentalists—people who engage in frequent pro-environmental behavior, including political and other collective actions—and compared them to people engaged in the fewest environmental behaviors, matching for demographic, educational, vocational, and socioeconomic variables. Interestingly, besides exposure to nature in childhood, the *loss* of a meaningful natural place by development was a common influence for environmental activists. The first inklings of eco-depression may spur some people to action. Influence by an author or book, and the influence of friends also played a role. Even people who didn't have childhood exposure were influenced by college or adult exposure to nature or conservation organizations. In maturity, some adults become environmentalists because of a transformative, often spiritual experience in nature. Overall, life history accounted for 56 percent of the variance between activists and those with environmental apathy. While this still doesn't totally resolve the "chicken and egg" causation issue, these results and others suggest that life experiences, including those at different ages, are reported by people as being instrumental in their own motivation toward environmental activism. Exposure to nature is one factor, along with the influences of family, friends, organizations, and even influential writers—the critical information link.[137]

Nonetheless, even for youngsters who have a love of the outdoors and wild nature, it's a fairly long cognitive leap from liking to be outside exploring green space to turning off the lights when you leave a room or taking shorter showers—some of the behaviors ascertained in studies as being pro-environmental that might correlate with nature affinity. That's even a leap for adults. It takes a lot of knowledge-based intellectual processing to see any obvious relationship between enjoying hiking in the woods and taking a short shower, or even spending time and money to put in insulation, or telling your boss you'll forgo that meeting because it's 2000 miles away and you think it's better for the environment to stay home.

But it might work. People who value nature tend to really value it. They are passionate about it. And passion can fuel these linkages. Cognitive connections are easier when the reward system actively promotes them. If nature becomes rewarding, as it has for many environmental advocates, that reward will be reactivated by defending and protecting and preserving the source of the reward. That's why the crayfish go to the part of the tank with the striped floor where they learned the reward would be, why songbirds learn how to stake out their territory, and why you'll try even harder when a mentor praises you.[138]

The Urgency for Behavior Change

In our current situation, the earth is the patient, and we are the guardians charged with decision making. The science tells us this situation is like an expanding blood clot on the brain before the patient loses all consciousness—it requires immediate attention. While we can't directly see the internal hemorrhage, *this is not just prevention.* The longer we delay, the worse the outcome, maybe to fatality. The sooner we act, the better the chances are of meaningful recovery—of maintaining some of the characteristics we recognize from our long life on our planet. Yes, our senses may have difficulty perceiving the problem, and we may not completely understand or trust the information provided by the experts, and a radical treatment like surgery is a frightening thing, with its own risks. It's not something anybody wants. But losing what we have is more frightening.

But here the neurosurgical analogy fails, because we can't hand the problem over to be fixed by someone else. The nature of this "wicked problem" is that we all need to fix it, including changing our own behavior and also becoming change agents for many other people, as advocates, policy supporters, informed consumers, knowledgeable citizens, workplace trend-setters, institutional leaders making tough decisions, political activists, and constant learners. We will need to call on all our brain capacities to override and overrule its tendency toward immediacy and concrete perceptions, and to proceed with patience and persistence, coupled with urgency. Perhaps some patience can come from recognizing the limitations and advantages of our shared but individual neural equipment, and using what we've got to move forward.

9

The Green Children's Hospital

IMAGINE YOU'RE A CHILD, and you're sick. You're told that to get better, you have to stay in the hospital. You get needles and tubes that hurt and tests inside dark blinking machines that make weird sounds; you don't know what's going to happen next. There's a TV in your room and sometimes you can go to a playroom with games and coloring. But the hospital is noisy all the time, people are talking about things you can't follow, it smells funny, and there are wires that attach you to screens that beep and the wires wrap around you and dig in when you try to move around or sleep. The sound of other kids crying upsets you. Everything looks shiny and hard. The place is full of strange things you've never seen before and don't understand.

Now imagine you're a child in a hospital bed. Your sheets have pictures of animals on them, and you face toward a window that pine branches touch, that you can open. There are birdhouses in the tree, and a bird feeder attached to the windowsill; the birds come right up to the window and look around. Out in the hallway the walls are covered with plants. Down the hall there's a glass tube with trees inside, and light comes from outside from way up at the top of the building. In the middle of the main lobby is a kind of a greenhouse; you can go in there with your family and make a wish by throwing pennies in the waterfall. There is something attached to your clothes that tells your nurses where you are and how you're doing when you're in the greenhouse or outside in the big garden, and your mom wears a necklace that lets the nurse call you to come back to the room if it's time for your medicine. When you're getting stronger, your therapist goes outside with you and you race to climb the little hill and jump off the log into the grass. You help plant some flowers in the greenhouse, along with another kid who's also in the hospital. On the roof is a garden where things you might have for dinner are growing. On one floor there's a big glass box with ants you can watch that make tunnels and carry leaves in and out of the passageways. After you're better, when you come back to visit your doctors, you can play outside and sit on the big rock that looks a little like a ship. If it's winter you can go in the greenhouse and see if the flowers you planted are still there. You pick up a sheet that you check off for each type of plant you find and get a prize when you find them all—it's your own little plant to take home. A sound through the necklace tells your parent when it's time to go to the doctors' office; you get to pick which animal makes the sound—a duck? An elephant? A frog? You don't like getting tests or X-rays, but there's a therapy dog you can visit when you're done, the same one you saw last time, who's really fuzzy and friendly. Your doctor gives your parent a prescription to take you outside along with a free pass to a local hiking area that is part of the hospital's program. There's a big board with pictures of kids who have gone there, and the nurse tells you that you can add your own picture next time you come.

Maybe this vision sounds quaint and out of touch. Don't kids want video games, action heroes and super-powers, interactive displays, pop stars, and sports figures?

But which is better for healing? For health? How would one find out?

Throughout this book we have explored how the reward system works, which behaviors matter most for environmental impact, and the principles of behavior change. We have assessed the evidence for biophilia, and the effect of nature on healing, restoration, and learning. We've learned that pro-environmental behavior requires alternate rewards to change deeply ingrained habits, and that it can be "nudged" by appealing to collateral benefits even without reference to an overt environmental agenda. We've seen what these kinds of changes look like at the macro, meso, and micro levels, and recognized the complexity of this challenge using our inherited decision-making tools. We've learned that no single approach will "fix" this grand challenge, and that all of us have a responsibility to act. It's time to test out these principles in a real-world example. What would we find if we tried to capitalize on the benefits of pro-environmental behavior in the form of a new "green children's hospital"?

In the United States, health care is an enormous enterprise, comprising 18 percent of the US gross domestic product. The goal of health care is to optimize health and well-being, decrease the rates of death, illness, and disability, and alleviate physical and mental suffering related to diseases and disorders. Practitioners of health care and the associated helping professions generally find their main mission—solving problems, helping people and making their lives better—*highly rewarding.* American medicine enjoys great success in this regard. Diseases eradicated, life expectancy prolonged, disabilities overcome, new scientific discoveries and technological advances all make medical care in the United States some of the best in the world.

And yet, the health-care industry is not exempt from our collective challenge to change our environmental trajectory. The health-care sector has a significant environmental impact, as research detailed later in this chapter has shown. And, as the population continues to expand rapidly, the health-care sector's reach continues to advance, catalyzed by public health successes in delivering these benefits to increasing numbers of people around the world. Short-term immediate health conflicts with long-term global sickness.

The green children's hospital idea was conceived as a test case, to investigate what the impact and obstacles would be of applying these principles—attempting to make pro-environmental behavior rewarding—to one particular part of daily life. It made sense to do this in the world with which I am most familiar, even though (and perhaps especially because) a link between "children's hospital" and "environmental decline" is not intuitively obvious to most people. That's precisely the point. Despite our inherent mission of helping children, health care generally has considered environmental issues and climate change as not "our shift." We are focused on a few children at a time, and mostly see our obligation to bringing all the resources we can muster to the specific child before us, rather than seeing that individual child in the larger context of the future of all children. Practitioners of academic medicine often get around this dilemma by participating in medical research—this way, you can try to make a contribution to the health of more children than just the few you are treating. But climate has not been a focus of most health-care practitioners; it seems too many steps removed from our immediate focus. What would happen if we tried to make this link between health care and the health of the planet that will affect our children's futures more explicit?

The connection between environmental issues and public health is not a novel one, especially with respect to things like air pollution and toxins in water and food. But an emphasis on a more direct causal link between climate change and public health is relatively recent. In this arena, because children bear the greatest present and future burden, pediatric health-care organizations have raised the first and most vocal alarms about the public health effects of global warming. The American Academy of Pediatrics was one of the first medical societies to publish a policy statement in its journal, *Pediatrics,* in 2015, stating this relationship in no uncertain terms:

> Climate change poses threats to human health, safety, and security, and children are uniquely vulnerable to these threats. The effects of climate change on child health include: physical and psychological sequelae of weather disasters; increased heat stress; decreased air quality; altered disease patterns of some climate-sensitive infections; and food, water, and nutrient insecurity in vulnerable regions. The social foundations of children's mental and physical health are threatened by the specter of far-reaching effects of unchecked climate change, including community and global instability, mass migrations, and increased conflict. Given this knowledge, failure to take prompt, substantive action would be an act of injustice to all children.[1]

The contradiction and tension between the urge to help one child at a time and the urge to direct resources to protecting the planet has accelerated in recent decades. The Academy's policy statement goes on to state that for those of us in health care, this is in fact "our problem":

> A paradigm shift in production and consumption of energy is both a necessity and an opportunity for major innovation, job creation, and significant, immediate associated health benefits. Pediatricians have a uniquely valuable role to play in the societal response to this global challenge.

So how might our vision of a "green children's hospital" fit into the big picture of climate change, the overall mission of health care, and what we can do to address environmental issues on an individual and organizational level?

There are data showing that people with children in the household have a higher level of environmental concern, perhaps reflecting the knowledge that climate change will affect them more than the adults; such may be the increased cognizance of pediatric health-care providers, teachers, and others dealing with children.[2] If we agree with the position of the American Academy of Pediatrics, there are certainly other things we can do about climate change as individuals, as we've discussed in prior chapters; we can vote for politicians who espouse progressive environmental policies, we can lobby our elected officials, we can vote with our pocketbooks regarding investments and product choices, we can change our personal lifestyle choices to reflect our belief in these priorities, we can join social and political movements. But what about our work lives, which often take up the vast majority of our waking hours? Can we who work in hospitals and health care better align our work lives with these particular pressing needs? The nurses and pediatricians who served in focus groups about this idea were universally on board—they all worry about climate change and would love to work in a facility that was showing the way forward in their work worlds as well as their private lives. Having this dimension would make their jobs even more rewarding, and boost morale. Patients and parents expressed enthusiasm as well.

As environmental writer Bill McKibben said back in 1998, while we should beware of people throughout history who have said that "the end is near" and "these are special times," in the case of climate change, accumulating data suggest that, this time, what we do in these early decades in the twenty-first century will, in fact, truly determine which way things go.[3] If we continue business as usual, things for the children we are treating or those of the same age in less fortunate parts of the world will not go well, as they mature and encounter the turn of the twenty-second century. So what difference can we make by changing the way we run hospitals—a little, a lot, or something in between?

What Do We Mean, "Green"?

"Green" has become a hackneyed term, applied to everything from household cleaners to companies that pump septic systems to whole cities, without any consistent standard by which to justify its use. You can't walk down the aisle of a grocery store without seeing in every category of product some that claim to be "eco" or "green"—but of course, this is a complex arena, and nearly impossible for the consumer to decipher (though there is some help available from on-line guides like DoneGood, Greenease, and B Corp). In an almost Orwellian twist of verbal appropriation, even corporations that sell petroleum duel with one another to apply the "green" label on some sliver of their own activities, like the paint they use on their tankers, thereby hoping to dissociate in the public's mind their corporate goals and the undeniable desecration of the natural world on which we all depend.[4]

When a small group of colleagues and I first set off to explore the feasibility of creating a new green children's hospital within my institution, the term was a bit less overused, and we had in mind several aspects of the word that might be applicable to a health-care facility. Our theoretical hospital would be "green" in several dimensions—all of which raised questions we would have to study. First, would it be feasible in a hospital to maximize energy and resource efficiency, to minimize consumption and waste, to reduce toxins, and to utilize the biophilic design principles that we saw were beneficial to health and well-being in Chapter 4? If it were in fact feasible to incorporate these principles into the energy-intensive health-care environment without compromising the delivery of high-quality care, would it make any significant impact from the environmental point of view, or would it be negligible, and not really worth the effort and expense? To proceed, this concept had to be more than just a marketing trick—it had to actually have a meaningful impact, not as an individual building (too small a drop in the ocean for that, obviously), but as a prototype and an approach that could be emulated in other health-care settings. If a new kind of hospital in fact had some reasonable impact environmentally, how then might we also draw people to such a project by appealing to the reward system, rather than because you "should" or "must" do something "good"? This is especially important if the project might involve self-denial (maybe the patients get the windows, instead of the doctors' offices) or prioritizing resources that might seem directed outside the institution's stated mission of providing health care to individual patients.

To attract people to this vision, we decided to explore using the "pull" of the inherent reward of exposure to nature—biophilia. If nothing else, our hospital would be perceived as beautiful, with the goal of adding a visceral appeal as well as intellectual and pro-social rewards to our enterprise. Coincidentally, about a year into our project, there had been a large public outcry about the destruction of a well-loved hospital garden at another local pediatric hospital to accommodate the construction of new buildings. The idea of a peaceful, natural space in a children's hospital seemed to resonate and be valued by the community more than might have been predicted on rational grounds. Full-page ads in the metropolitan newspaper protested the decision to bulldoze the garden. As the neuroeconomics lessons in Chapter 3 revealed, the human decision-making apparatus does not rest on logic alone.

In our initial phase exploring feasibility, our small team planned to gather facts to learn what it would take financially, what effect it would have from an environmental point of view, what the effect on health and well-being of patients and employees might be, how the reward system of caregivers might be affected, as well as what we could learn about the practical and behavior-related barriers if we were to create an experimental green children's hospital as a case study. What we learned quite early along the way was that we had to employ these principles—what different decision makers would find rewarding—just to get the project initiated and off the ground.

Investigating the Evidence

The first step was to answer some of the basic factual questions. If the idea made no sense from an environmental impact point of view or was prohibitively expensive, it would fail before it started, and might be viewed negatively as a misguided marketing campaign aimed at a quixotically small segment of our population for whom this theme might resonate. Additionally, the idea needed to be seen in the context of a specific institution. At the time it was conceived, my hospital, a major academic general hospital serving patients of all ages, did not have a dedicated space in which all pediatric efforts were concentrated. While many of the pediatric providers felt we were behind the times facilities-wise as an identifiable children's hospital, we still had been able to recruit excellent people and shared outstanding technical resources with our adult colleagues, and so could deliver generally excellent care, despite our physical challenges. Many practitioners who cared for children had long felt that having a dedicated space would be advantageous for improving the efficiency of care, communication among providers and teams, standardization of pediatric-oriented processes, reduction of redundancy and risk of errors, and also would greatly improve the patient and family experience. But it was also true that building or renovating a space to create a separate children's hospital on our campus was not at the top of the priority list for the institution overall.

Nonetheless, leadership had made it clear that they wanted to continue to care for children at the highest level of excellence for the foreseeable future, notwithstanding that we were located in a city with several other major pediatric facilities. So we had a number of hurdles to overcome, and the "green" aspect might complicate—or help—our cause.

The first question was whether making a hospital more environmentally responsible was even worth it from an impact point of view. Its primary mission is making people well, and this is an energy-intensive enterprise. And an early realization at the inception of the project was that when you start talking about hospitals and the environment to people who work there, the first thing they think about is *waste.* What we throw away in an operating room or procedure suite, compared to what we use, is enormous. So much of what we use once and discard is plastic—from syringes to intravenous bags to bedpans to disposable scalpel handles to plastic trays for kits of supplies to caps on medication vials. The list is endless. When you think about medical waste washing up on beaches, it's all this plastic stuff.

But this is another manifestation of the brain-based tendencies explored in prior chapters—we tend to focus on what we can see, on what is tangible. In terms of carbon footprint, isn't there likely to be an even greater effect of other things that make a hospital run? How much fossil fuel do we use, both for the energy to heat and cool our buildings and run our machines, and what about the almost unfathomable life-cycle considerations for all the stuff we use, all the medicines we give, all the equipment we rely on to do our job of healing? Can any of this be changed in a way that doesn't interfere with the mission to care for patients?

As it turns out, these environmental costs are not insignificant in the big picture. In the United States, health care's percentage of the US gross domestic product is probably the highest in the world. This high number results from a variety of factors and has been analyzed from many directions. One consequence of the size of this enterprise is that the heath-care sector—arguably the "good guys" in the eyes of those who need it—is also a major contributor to the carbon load of the world's historically biggest carbon producer, the United States. So this required a look at what health care in general, and hospitals in particular, contribute to environmental impact in terms of energy, waste, and toxins, and also ways in which hospitals could serve as laboratories for these concerns and other important elements with environmental impact, such as food sourcing.

Energy and Other Environmental Impacts

The extent and manner in which the health-care industry contributes to carbon emissions and other environmental stress has been addressed by a variety of research approaches. One common method for estimating environmental impact of any product is called "life cycle analysis." We first encountered this concept in Chapter 6, as researchers calculate the impacts of various consumer choices. In this type of assessment, the impact of all components used in a specific process—say, building a new sofa—is taken into account. Where was the wood for the frame grown, and was it replaced with new planting? This is important to maintain vegetation that will continue to absorb and store carbon, so as not to create a net loss of this balancing capacity, by, say, cutting down trees for lumber and then paving over the land and building a warehouse. How was it transported to the place where the furniture was manufactured, and how much fuel did that require? What about the fabric, the foam, and the springs? What kind of energy was utilized for manufacturing, and what kinds of toxins arise from each step of the process from raw material to finished product? Are there flame retardants, and materials that give off volatile organic compounds? How long is the sofa expected to last, and what will become of it when it has worn out? If it goes into landfill, are there things that will leach out? This type of analysis is sometimes called "cradle-to-grave" assessment, and it can be used to compare the environmental impact of specific types of goods or of entire industries.

In 2009, the *Journal of the American Medical Association* published a study by researchers at the University of Chicago which found that the health-care sector accounted for 8 percent of all greenhouse gas emissions in the United States.[5] To calculate this number, the researchers used a life-cycle assessment approach. When applied to health care, this methodology analyzes the impact of each commodity purchased by health-care institutions, accounting for all greenhouse gases produced both indirectly by the supplier and directly by the health-care entity. The results are then translated into units of million metric tons of carbon dioxide equivalent, which allows for comparisons among industries. Interestingly, approximately 80 percent of the total global warming potential from this sector was due to carbon dioxide emissions, and about half of the carbon equivalents in health care arose from direct activities, meaning things that happened on site. Among the subcategories of health care, hospitals were the single largest contributor, accounting for almost 40 percent of the industry's contribution. A different group of researchers looking at health-care expenditures over a ten-year period found similar results, concluding that the health-care sector contributes 12 percent of the total acid rain burden in the United States; about 10 percent of greenhouse gases, smog, and other air pollutants; and also significant amounts of other toxins.[6] These amounts have been increasing over time. As in the first study, the single biggest component was hospitals, and the single biggest activity was energy usage.

Health care is in the business of improving health, but how do these greenhouse gases and pollutants associated with the health-care industry impact health itself? To compare impact on health from various causes, researchers often use a metric called disability-adjusted life years, or DALYs—that estimates how much health is lost due to a specific cause. The health effects of health-care sector emissions and toxins were calculated to cause as much impact on health as do all preventable medical errors, which have received much more public attention and lead to tens of thousands of deaths in the United States per year.[7] This means that reducing the direct and indirect carbon emissions associated with hospitals is a goal that could have a significant impact on the overall environmental big picture—and on the health of the population.

Waste

Hospitals have been estimated to produce five million tons of waste annually, 6,600 tons of waste every day collectively, or approximately 29 pounds of waste per day for each staffed hospital bed.[8] If several layers of packing material bother you when you order a product in the mail, what happens in hospitals seems like an order of magnitude worse. Why so much? Because everything you use in a hospital is new, and just for you. Everything that is sterile is packed in layers on layers of impermeable plastic and other kinds of synthetic non-biodegradable refuse. Every medicine comes in a package that is thrown away. Every vial, every syringe, every intravenous line, every bedpan, every identification bracelet, every bandage. The operating room alone creates about a quarter of all the waste, because of even more packaging, sterilization wrap, and disposables. Food waste, toxic waste, pharmaceuticals—all the unused medications that need to get thrown away just so—it's an enormous amount. Waste stream expenses make up a good part of hospital budgets.[9]

Hospitals can do better, and often can save money by doing so.[10] The barriers are low motivation at the top, and the challenge of changing habits in the workforce. In the operating room, unless non-biohazard waste trash bags are plentiful, people will tend to just throw everything in the "red bags"—biohazard, or "regulated medical waste"—just because those are the ones that are there. Most of the waste in an operating room is not hazardous at all, much of it can be recycled with some creativity using "circular economy" principles, and disposal of red-bag waste is nearly ten times more expensive than standard, "clear bag" trash.[11] Hospitals that successfully teach—or, more successfully, nudge—people to appropriately sort the trash into true biohazard waste, recyclable waste, and ordinary trash destined for the landfill have saved significant amounts of money.[12]

The savings is the main incentive for the hospital; it's a bit more effort for the workers, who don't typically realize the savings directly, but maybe feel better about doing their part to help. But in the operating room, efficiency and getting the job done quickly and smoothly are prioritized. When they unwrap something—and almost every minute, someone is unwrapping something—the nurses and personnel have to stop and think: Where does this stuff go? Is it on the list of things that we recycle? Is it regular trash? Does this bucket full of stuff go into the red bag, or the regular? Now, these are altruistic people, but recycling is not what they signed up for; it may seem a bit removed, even with a "save money" argument. Habits are hard to change, especially when the reason is—as we've seen—remote, removed from your experience, communicated by someone you don't know, and doesn't benefit you directly, but adds to your cognitive and physical workload just enough to be perceived as annoying. Taking lessons from other aspects of behavior change strategies that we saw in Chapters 7 and 8, what might work better—competitions among operating room teams where the workers doing the sorting get a financial bonus? Getting them invested in process improvement, and having teams of their most respected peers work out how to best accomplish this, with their input? Making the environmental benefits a part of their mission, and part of what gives them pride in their work? Maybe which combination of strategies works best, and how much it saves, would be a research project for our "living laboratory" Green Children's Hospital. Once we learn, we can apply it to the larger hospital system throughout our network, and save even more—money, and landfill space.

A more complex job involves working with the outside suppliers to reduce the waste in the first place, at least in packaging—but the incentives have to be there for them to change; and changes must not hurt the suppliers financially.

In this kind of effort, collective action through participation with organizations like Practice Greenhealth and group purchasing organizations that join hospitals together to create a larger impact in purchasing decisions can increase the effectiveness of this life cycle approach. We will learn more about those important organizations later in this chapter.

Green Buildings, Indoor Air Quality, and Health

Have you ever had the experience of purchasing a new piece of furniture, and you get it home and notice that it smells awful? Not awful in the something rotten way, but in the acrid, chemical way? Or new carpet is installed in your office, and you notice that your eyes are watering? Volatile organic compounds (VOCs) are chemicals that are emitted from a variety of indoor furnishings and surfaces. "Volatile" means that the chemical can disperse through the air and be inhaled by people. One example is formaldehyde, used in the medical world to preserve tissues in jars, and cadavers. Examples of VOCs include solvents and cleaning solutions, air fresheners and scents, substances in building materials and adhesives, dry cleaning chemicals, flame retardants, preservatives, stain-resistant coatings, and office supplies like copier fluids.[13] Some materials also shed particles that can be inhaled, like fiberglass and drywall dust. There are microbes—fungus and bacteria that can live and reproduce in air ducts and moist spaces. Many things concentrated in buildings are known to be harmful to human health, increasing the risk of respiratory illness, infections, and cancer.

The history of indoor air quality parallels changes in our human circumstances and technologies over recent time. Until the past several decades, dwelling places were not even close to being airtight. Throughout history there were gaps in the tent flap and hut, openings for smoke to escape and outside air to get in. Even wooden or stone houses had gaps at rooflines and around windows and doors. Anybody who's been in a drafty old house knows this. But when oil prices rose precipitously in the 1970s and 1980s, a focus on energy conservation led to stricter building practices designed to make houses and office buildings more tightly sealed. This led to a reduction in air-exchange rates, sometimes by tenfold, and to beneficial reductions in energy waste in the form of heating or cooling.

Shortly thereafter, health-care providers and epidemiologists began to notice clusters of office workers with a variety of overlapping symptoms. These most often included headache, fatigue, difficulty concentrating, eye or throat irritation, cough, skin dryness or rash, and often a perception of unpleasant odors.[14] This phenomenon was termed "sick building syndrome," and it has been the focus of considerable research and regulation. National and international standards have been enacted to limit levels of dozens of different kinds of noxious substances in the indoor environment.[15] New ones pop up in specific circumstances—in hospitals, these include cautery smoke from surgery in operating rooms, heliport fumes from the Medflight helicopters, air laden with the scent of rat cages, bacteriology culture plates, formaldehyde, anesthetic gases.[16] A subcategory of sick building syndrome is called "sick hospital syndrome."

To counteract this trend, new certifications and standards were designed to try to promote energy efficiency without exacerbating sick building syndrome. Smoking was prohibited in most public buildings, and air quality standards such as those of ASHRAE (American Society of Heating, Refrigeration, and Air-Conditioning Engineers) or the United States Green Building Council LEED (Leadership in Energy and Environmental Design) certification, became more widely adopted. The mainstay of improved indoor air quality is a two-pronged approach to decrease or confine the use of materials and substances that contain hazardous materials, and to increase ventilation rates.[17]

But like those around the earth itself, some substances that we can't perceive as irritating or harmful also can have deleterious consequences. Carbon dioxide is sedating. Its relative, carbon monoxide, is lethal. Both volatile organic compounds and increased levels of carbon dioxide—what you get in a crowded conference room, or an elevator or airplane that feels "stuffy"—*can decrease your cognitive function.* In one set of carefully controlled studies, researchers used an experimental office building in which the indoor air quality could be manipulated from the floor below, without the occupants' knowledge of what kind of air they were breathing. With informed consent, and using air constituents within generally recommended levels, volunteers performed their usual professional-level work in the offices for the duration of a typical workday. At the end of the day, they were given detailed cognitive function tests designed to reflect real-world professional cognitive demands and decision making, including measurement of errors. The results? Subjects who worked in air with the lowest (healthiest) ranges of carbon dioxide or volatile organic compounds, at levels that are required for the highest level "green building" certification, performed significantly better—with scores over 100 percent higher—than those exposed to the more concentrated "conventional building" range of these substances.[18] The conclusion is that greener buildings that focus on indoor air quality improve both the health and also, unexpectedly, the productivity of their workers. Since worker sick days and productivity are some of the major determinants of the cost of operating a business, these types of data support a strong economic argument for the relatively small up-front cost of the best green building design.

In the specific context of a children's hospital, both staff and patients benefit using both health and economic metrics, as has been shown when hospitals convert to green design.[19] Staff and family satisfaction increase, staff turnover and sick days decrease, and patient infections and even mortality go down. This provides the most powerful incentive for making change.

Food, Health, and Environment

At one children's hospital where I worked, one of the attractions was a fast-food hamburger franchise right in the lobby. The word among the staff was that this was one of the highest grossing outlets in the country. The kids loved it, parents used the treats as a bribe for good behavior in the waiting rooms or x-ray suite, and the aroma of French fries masked the usual "hospital smell." Even the surgical team could dash down and get a burger before an operation. Eventually someone solved the mystery of strange densities noticed by the radiologists that appeared on the abdominal films of many children treated there: they were thought to be tiny bits of calcified bone fragments in the burgers.

Things have changed some since then. As we've seen, agriculture and land use are major contributors to climate change. The typical American diet is a major contributor to health issues. Many authors have produced evidence that red meat production is one of the big culprits in environmental decline.[20] In a fortuitous alignment, food that is better for the planet generally also is better for health. Diets in which plant-based food plays the major role are healthier, and often can be grown and produced locally and with low toxicity.[21] While there may still be a fast food outlet near that hospital, it is no longer in the lobby.

One would think that hospitals naturally would focus on healthy food for patients, families, and staff. It is true that patient diets are carefully monitored by hospital dieticians, but the environmental side of the equation is not often considered. This has been changing some under the influence of organizations that sponsor initiatives whereby hospitals can pledge to improve the environmental impact of their food services. Practice Greenhealth, for example, is an international organization that promotes sustainability in many facets of the health-care sector. It provides specific suggestions regarding protein sources and other nutrients optimized for health and climate, while also facilitating group purchasing options for its members.[22] More than 1,000 of the 5,700 hospitals in the United States participate through a variety of organizations in initiatives to promote healthy and sustainable food services, and many participate in group purchasing organizations that enable institutions like schools and hospitals to combine their purchasing power to facilitate the procurement of food with these qualities.[23] However, like other dining services for "captured populations" such as schools, if people don't like the food, it will go to waste; this doesn't improve sustainability. Hospital food service is a complex profession. Tastes change slowly, and hospitalized patients are not typically in an adaptable mood. If they get white bread at home, whole wheat may not go over well. If a slice of pizza seems appealing, a cabbage and beet salad with quinoa may not be greeted with enthusiasm. Illness is not usually the time people have the most reserve to try something new. It's called "comfort food" for a reason. Remember our rats in Chapter 3; under stress, the brain prioritizes familiarity over novelty. The same principles likely apply to hospital visitors, who are stressed and rushed.

Employees may be a different situation. These folks know something about nutrition and might welcome more healthy choices at their workplace. (As long as it tastes good, isn't too "weird," is quick and convenient, and not too pricey.) But environmental concerns? Maybe, but the health aspect would seem a more logical priority. So food with a better environmental profile is not going to be something that likely will drive purchasing unless there is some external reason—better price through a cooperative, produced on site, or for public relations. Research into institutional food and the environment is in its early phases, and much still needs to be learned about possible synergies with health, health care, and food—but the food has to be affordable, and something that people will eat and not send back to the kitchen with complaints. In our green children's hospital, we could study what works best.

Cost

The cost of building a building is only the first step; it's the ongoing operations that count the most over time. We already knew that compared to other hospital operations, pediatric care delivered a lower profit margin (more on this issue below). What are the facts on the effect of ongoing operations in buildings like the one we had in mind?

When green buildings first began to be promoted, with features such as better insulation and improved energy efficiency, maybe even with some alternate energy sources such as solar, it was clear that they would cost significantly more up front. The hope was that the owner would recoup the loss by energy savings over many years, typically measured in decades. With respect to institutions, as in politics, if you're a leader making such an investment decision, you likely will not be rewarded when the books become unbalanced during your tenure, despite the hope that they'll balance out in the future, when it's someone else's turn to reap what you've sown. Your board of trustees is judging you on your own balance sheet, not a theoretical one decades away. Humans are short-term evaluators. We struggle with long term. This is your brain design at work.

During the past two decades, a new movement in architectural and building philosophy began to develop. Interestingly, this was based in part on a similar movement in the medical field. In the health-care sphere, historically the choice of treatment for a given condition rested on the experience of individual practitioners. But with the growth in access to more objective and reliable "big data," partly reflecting the increasing use of electronic health records and large insurance and other administrative databases, along with the computational tools to analyze them, researchers had new opportunities to look at the outcomes for large numbers of patients to see what treatments worked best. This approach is termed "evidence-based medicine," and while it has its proponents and its detractors, the use of evidence for treatment decisions, tracking of outcomes, and even reimbursement has become widespread in health care.

Along similar lines, recent decades have seen the rise of a concept in architecture called "evidence-based design." Rather than just opine that biophilic design makes people happier or more productive, and that energy savings help reduce operational costs, and that green building techniques reduce waste and toxicity, architects and institutions can collect data and use it to inform future choices. One can argue that the 1984 study of Roger Ulrich that we examined in Chapter 4, showing that hospital patients looking at trees recovered faster than patients looking at a brick wall, was the first demonstration that the design of your hospital and the incorporation of nature could affect your outcome—and how much your stay costs the hospital.[24]

In 2000, a group of health-care institutions and designers came together to begin to create case studies using this approach. Termed "pebble projects," for the "ripples" they might create, these case studies analyzed what made hospitals more effective, safer, less wasteful, and more cost effective. Ultimately the collaborators presented compelling data that there was a clear business case to be made for better-designed newly constructed hospitals, renovations, and additions that met the dual goals of better health and better environmental performance, and that achieving both goals correlated with substantially improved return on investment over short time frames.[25]

Data from these and hundreds of other studies have shown that features like single-patient rooms, noise-reducing materials, improved ventilation, logically placed hand-cleansing facilities, floor plans that facilitate nursing efficiency and patient communication, and other evidence-based design elements have been demonstrated to make both patients and staff healthier and happier. The approximately 5 percent increase in up-front costs is estimated to be recouped within the first year, and the savings recur each subsequent year.[26] Like the field of medicine, architecture becomes a laboratory to test what works best and refine it to work even better.

What about the cost aspect of healthier food? Most people know that organic and fresh foods come at a financial premium. Some of the premium for quality and freshness can be offset by the fact that in general, plant-based foods are less expensive than meats. Reducing the amount of meat in dishes served and increasing fruits and vegetables is in alignment with World Health Organization recommendations for health, delivers a more favorable environmental profile, and can save money. So-called "values-based supply chains" have been instituted by networks of hospitals, working with and outside of traditional group purchasing organizations, to reduce the cost of healthier and more environmentally responsible food in the health-care setting.[27] When large corporations like Walmart bring their enormous purchasing power to the table, they can set the stage for growers to comply with the purchasers' demand for locally produced food grown with health and sustainability in mind. Hospital networks may have smaller but still potent power. However, some analysts warn that this may undermine the sense of place and local community that the local food movement at its best is able to incorporate.[28] You can't get any more local than one hospital in our city got when it developed a large roof garden for fresh produce, using half for the hospital and donating half to the local community. In a hospital setting, adding this kind of endeavor to the sense of mission and advocacy for the local community may add value to the proposition.

Of note, some of these financial gains from evidence-based design are predicated in part on a patient population that can expand the hospital's business. In our specific setting, with an overall regional pediatric population that wasn't projected to increase, this meant diverting patients who currently obtained health care at competing centers. Our business case would be more difficult, with few data on which to base predictions. Kids make a lower margin than do adults at our center, and somewhat paradoxically, we happen to exist in an intensely competitive environment for pediatric care. However, even apart from growth in patient volume, case studies on improvements in energy efficiency, waste reduction, and operating room supply efficiencies have demonstrated such favorable returns on investment that they have been recommended as ways to help "bend the health-care cost curve" in the increasingly challenging landscape of medical reimbursement, just because of increased savings on operations.[29] Metrics and checklists have been designed to help hospitals travel down the path of greening health care; a growing dataset is available to serve as a base for this effort.[30] If we were able to cut costs and also attract new patients and programs because our new pediatric hospital was both healthier and provided better outcomes, and because it was beautiful and incorporated new technologies that improved the patient experience, the combination of rewards could be a winning formula. If this "green" theme could further enhance our existing partnerships with organizations like the Appalachian Mountain Club, whose "Outdoors Rx" program teams with our pediatricians to get kids into the outdoors, this would be good for the lifelong health of our community. In addition, through our function as a "living laboratory," capitalizing on our rich interdisciplinary academic collaborations across departments and schools within our affiliates, we could make unique contributions to the rest of our center and to other children's hospitals that could learn from our efforts, and this would be a source of pride and identity.

At this point in our story of this project, now we've got a proposal for a healthier hospital, with better indoor air quality, less waste, better energy efficiency, healthier food, and less employee turnover. All that sounds great, but the next challenge is to make the benefits of this hospital with environmentally optimized features attractive to hospital decision makers and possible users of the facility *in the moment*, not in some theoretical future. As we've seen, our brains are better able to perceive the value of something we get right now, rather than something far away in time. This is where the lessons of biophilia and biophilic design can be put to the test. This hospital has to be beautiful. It has to be a *destination*. We have to find ways to embody the theme of nature and the environment as part of a healing experience, from newborns to teens and young adults, along with their families. It can't seem boring or preachy or irrelevant. We have to find ways to make the staff feel that their workspace respects their work, cares about them, and isn't adding "fluff" or silliness that makes the expert application of their skills more difficult. Acute care medical practice relies on intense, critically demanding human processes and sophisticated technologies, and we can't hinder those requirements. We have to do "green" in ways that benefit everyone, in a location in the middle of a northern city, in an energy- and technology-heavy enterprise. This will be a challenge, and we will need help.

Germination

As I learned more of these facts, and in the context of the conviction shared by many pediatric providers at our hospital that we needed a new pediatric facility to provide the best care and to stay competitive, I began to seek input from a variety of stakeholders. Within and outside our own organization, we needed to learn whether and in what ways we could help people decide to change at the different levels that influence organizational decisions. These included hospital leadership, chairs of the various specialties, practicing physicians and nurses, parents and community members. The first reaction most people showed was puzzlement—how did "green" connect to "health care," "hospital," and "children"? But there was enough encouragement from some key people to get the project into the germination phase. One specialty chairperson told me, "I don't agree with your politics, but as a marketing strategy, it's brilliant!" This, then, was the honeybee strategy discussed in Chapter 8—even people who didn't agree that environmental goals were important could support the idea of a novel and distinguishing identity for our pediatric facility.

The most important early advice from hospital leadership was to talk to architect John Messervy, who was the Director of Capital and Facility Planning for the network of university teaching hospitals to which my hospital belonged. John was an experienced and highly respected leader in health-care building projects, and was in some capacity involved in of all of them for our entire network. I did the minimum homework into John's background—not enough, as it would turn out—and was a bit surprised that he and his associate, senior architect Hubert Murray, even agreed to meet with me. It was with some trepidation that I ventured to corporate headquarters in an unfamiliar building a few blocks away from the hospital. We sat around a table, and I introduced my idea of the green children's hospital—where the idea arose, why it might be a novel approach and worthwhile, how it could be a prototype, a laboratory to test new ways to reduce the environmental impact of health care, and in a setting designed to incorporate aspects of nature known to be good for healing and reducing stress. How this could distinguish our small pediatric hospital, be cost effective, and be a novel contribution to the future of the children we treat. John listened, not saying much, and giving little away by his expression. I couldn't tell during the presentation if he thought I was a nutty clinician with ideas that made no sense, totally out of my league, wasting his time, or really how he was reacting.

I finished and waited, through a pause that seemed minutes long. Then John nodded slightly, as did Hubert, more animatedly. And John said, "That's exactly what we should be doing, and we should be doing more of it." For a moment I thought maybe I had mis-heard, but no, he meant it. I asked again to be sure. The homework I hadn't done would have revealed that John already had won major awards for sustainable design in health care. This was groundbreaking news, and while I knew we still had major hurdles to overcome, this was at least one powerful person who agreed in principle that the idea had merit.

Building a Team

With John's help, the next stop was Gary Cohen, founder and CEO of Healthcare Without Harm. Gary had gotten his start in the toxicology world, and he had discovered that chemicals in the plastic bags and tubing used to deliver intravenous fluids leached into the bloodstreams of newborn babies in neonatal intensive care units. In fact, my hospital had been one of the test sites for this study. Another hospital in our network was another, and when that hospital switched IV bags and tubing before ours did, our babies continued to have high blood levels of the chemicals, while theirs went down. Eventually, switching to the less harmful bags became the standard of care, and Gary grew this discovery into an international organization dedicated to reducing toxicity in health care. Health Care Without Harm also designed ways to enable hospitals to improve their environmental profiles by banding together in group purchasing contracts through suppliers vetted by Gary's team of experts for environmentally sensitive and low-toxicity products, through their affiliated organization that was mentioned previously, Practice Greenhealth. This collaborative mechanism provided member hospitals with a potential means to overcome some of the "it costs more" hurdle of healthier, greener supplies. Despite an enormously busy schedule, Gary became an important cheerleader for our project, pointing out that another building in our system shepherded by John Messervy, a new rehabilitation hospital right on the water in the Charlestown Navy Yard area of Boston, had been built to be resistant to sea level rise and climate change. Gary believed that our proposed children's hospital could become a similar beacon of cutting-edge environmentalism in the health-care world, which was exactly our intent.

What I learned, to my surprise and that of many clinicians I asked, was that my hospital individually, and the health-care network as a whole, already had signed on to some of Practice Greenhealth's hospital initiatives. These included commitments about energy efficiency, reducing waste, and encouraging local and sustainable food purchasing. The plastic utensils and Styrofoam dishes provided by the very nice people at Food Services for every function that served coffee and donuts or sandwich wraps did not seem like things that came from a "sustainability" commitment. Our green children's hospital project might be an opportunity to learn the best ways to both look and act the part of an institution committed to these principles, by means that the staff and the general public could see in plain sight.

As the project progressed, people who heard about it started to send me suggestions. One was to connect with people at the T. H. Chan Harvard School of Public Health, who were interested in healthy buildings as well as biophilic design. Two stars there were Julia Africa and Joe Allen. Julia had run the program for Nature, Health, and the Built Environment, had been the first author of a widely cited reference on this topic, and at our first meeting taught me that Erich Fromm was the often-uncredited source of the term "biophilia."[31] Julia had valuable experience and insights into the rationale for and history of biophilic design. She also had a practical knowledge of the challenges in urban environments, and what could work to bring the most important elements to a project. She knew the most effective ways to put people at ease in a space. And she had done some detective work on what kinds of things have been tried successfully in hospital environments to overcome some of the tricky issues like pollen and mold when bringing live plants into patient areas. She also understood why this concept might be of particular relevance to children with illness and their families. Children, she said, don't just have to manage their own fears; they often "manage up," and try to protect their parents from worrying about them. A soothing nature-filled place could help both parent and child deal with some of this stress. But she also knew the practical issues: that bird guano was destructive to buildings, for instance, so might temper my idea of installing bird feeders outside the windows. We joked a bit morbidly about what if a raptor—they populate cities now—attacked a bird and traumatized the kid watching out the window. Julia became a trusted sounding board and resource, connected to many useful people and research findings in this area.

Joe Allen ran the University's Healthy Buildings program, and he had developed multimodality sensors to pick up all kinds of metrics about indoor air quality. His research showed that something simple, like the carbon dioxide level in a room, affected worker productivity significantly; this is why people often get sleepy in airplanes or stuffy conference rooms. For a small investment in air circulation, a business could improve the cognitive function and decision-making capacity of its staff to create a very favorable balance sheet by improving worker performance, as well as creating a healthier indoor environment. That small changes might make a significant impact also was true for volatile organic compounds and other chemicals often found in furniture, carpets, and other materials in closed spaces—his students had looked around the college campus and found these substances everywhere. Such chemicals can be toxic to humans; having a way to keep them out of the hospital environment would make for healthier patients, families, and staff. We began to see this project as something that might serve as a prototype not just for energy efficiency and waste management, but also for finding innovative ways to improve the healthiness of the building itself.

Other important partnerships arose around the engineering challenges this facility might encounter. Dr. Mary Tolikas was at that time the Operations Director at Harvard's Wyss Institute for Biologically Inspired Engineering, and as an engineer and technology maven, she was a powerhouse of information and enthusiasm about collaborative opportunities. The idea at Wyss was that nature itself serves as the inspiration for engineering advances to solve problems, many with medical applications. We brainstormed about two possible projects as pilots. The first involved an ultra-smooth surface that Wyss engineers had developed; it was so smooth that bacteria couldn't adhere to it. Might this material be useful in a hospital environment where infection control was a constant (and expensive) battle? And would cleaning such an ultra-smooth surface require fewer toxic chemicals?

Another challenge that might be amenable to a Wyss solution was the problem of plants. At my hospital, indoor planters lining the window wells in some of the corridors ultimately had to be removed because of tiny mites and mold. Others hanging from a brand-new two-story atrium of our latest and "greenest" building didn't survive the harsh light through the floor-to-ceiling windows and were replaced with (gasp) plastic plants, that smelled like, well, plastic. To my mind, this experience was worse than if they had replaced them with shiny mobiles, like those in the famous big children's hospital I worked in previously. There the lobby had been filled with trees in planters, and the kids ran around among them and climbed on carpeted steps down to an ameboid wishing pond, into which they threw dimes and quarters begged from their parents. Their laughter filled the whole open atrium, and along with the sound of the running water in the fountain was pleasant and made us smile. But with the last renovation, they replaced the trees and the fountain with one of those Rube Goldberg machines encased in plexiglass, that ding dinged incessantly, driving everyone nuts. The carpets gave way to polished, echoey granite surfaces. They hung huge metal mobiles that were futuristic and space-like, and we all wondered how they would dust them up so high. Over time, it became clear that dusting did indeed prove a challenge.

So how to manage plants in a hospital? We talked to the engineers at Wyss about some kind of translucent membrane, with which we might create a greenhouse without glass. I was inspired by the botanical greenhouse at Smith College. The students there often took "green breaks" during the harsh western Massachusetts winters, wandering through the ferns and dripping humidity and just soaking up some kind of healing balm. Many credited the greenhouse with getting them through the long bleak period after the December break, when spring was months away, and the workload and stress often peaked. Couldn't we create something like that for our kids and parents, and still protect those with allergies or sensitivities by encasing it in something that would let the sound and smell and moisture through, but keep the potentially harmful substances confined? Wyss scientists thought this was an interesting materials engineering challenge, and they agreed to help.

Up to this point I hadn't thought much about food and hadn't perceived a need for a team of collaborators for this topic. Several things changed my mind to make it a priority, as noted in the overview earlier in this chapter about how we envisioned our new prototype hospital would work. First, Michael Pollan, author of *The Omnivore's Dilemma, Cooked,* and other popular books about food and the environment, was a Radcliffe fellow at the same time I was and had an office two doors down from mine. Michael raised my consciousness about this issue, as did an incoming fellow the next year, Gideon Eshel, who studied the environmental effects of agriculture. These people and others in this area had shown that there were better ways of doing things that were good for health, better for the environment, and importantly, did not have to be prohibitively expensive.

Another source of food-related information came from the Practice Greenhealth efforts. This group also promoted healthy food practices for hospitals, and ours had committed to using locally sourced and healthy food in its operation. I had noticed the "dot system" labels in the cafeteria with a "red, yellow, green" scheme, which ranked choices that were better for health. But what about the environmental impact?

Around this time, I was approached by an early career researcher who was about to finish a doctorate in nutrition. Stacy Blondin was planning a postdoctoral project using food labels in the dining service facilities at the college that would designate both the health aspects and the environmental impact profiles of food choices. Food labels regarding environmental considerations might influence choices at a captive-audience, progressive place like the university campus, but would that translate to a facility utilized by the general public? And if your sick kid wants pizza and French fries, what parent would say, "No, dear, it's kale and tofu for you today!" That's just not how the brain reward system of a parent likely works. But this was something we could study. Even for parents, the immediate wants and needs of the child before them likely outweigh a theoretical future of progressive global warming and food scarcity. But what if we had a roof garden, where the kids could participate in the growing of healthy food? As the savvy reporter Shira Springer noted in an article profiling our green hospital project for the Boston Globe, "After all, what kid doesn't like to play in the dirt." Boston Medical Center had done this with great early success, providing for the hospital as well as the local community. I signed on to Stacy's food labeling project with enthusiasm, with the understanding that we might apply what was learned to our green children's hospital.

Funding

By this time, hospital leadership had approved supporting my time on the project for one day a week, at a standard research salary level. There were also some small funds available for planning meetings that could bring together local and national experts to think through feasibility and logistics. This was a major victory and conferred credibility on the project. Being a smaller children's hospital within a larger general hospital environment had some advantages—we wanted to create a prototype, and doing this on a smaller scale would be a lot easier than if we were a very large children's hospital trying to go "comprehensively green" within an established infrastructure and culture.

We were still a long way from securing a commitment to go forward with an actual building. But now that we were allowed to talk about the feasibility stage of the project publicly, our next goal was to formalize some research that would provide data as we tried to gain support for the project through the various layers of institutional leadership. This research had two purposes; we actually wanted to learn what might work best and share these findings, in typical academic fashion, an approach solidly in our comfort zone. And we still had a lot of work to do to convince those in decision-making roles—hospital administrative and financial leaders and trustees—that this project should be undertaken. Our case would focus on several dimensions: effect on the health of patients and staff, cost effectiveness, innovation, community benefits, public profile, and alignment with the overall history and mission of the institution.

To this end, we applied for a grant from the university's Climate Change Solutions Fund. This funding source awards financial support to several environmental research projects each year. We emphasized the "hospital as a laboratory" angle—something that would continuously test and refine how to meet environmental goals within the health-care sector, findings that could be shared with other hospitals, thus magnifying the potential reach of the endeavor. We included projects on modeling optimal energy efficiency delivery in our urban space—conservation, solar, geothermal, battery, or combinations of these sources. One project would test indoor air quality with Joe Allen's sensors in some of our older, existing spaces for kids compared to some of the newer, greener building projects overseen by John Messervy's innovations in our health-care system. Another research project focused on the psychological effects of exposure to green space on nurses by randomizing break time to our oncology healing garden or a standard conference room. Would this brief an exposure to nature make them less stressed, and reduce burn-out and turnover? The Wyss Institute projects involved testing the ultra-slippery surfaces in our pediatric operating room, measuring bacteria counts on this surface compared to standard stainless steel at the end of an operation, and determining whether the slippery surface was easier to clean and could lead to a reduced need for toxic cleaning processes. Finally, Stacy Blondin's "food team" would do a parallel study in one of our hospital cafeterias, looking at the effects of both nutritional and environmental labeling of foods on choices made by visitors. This would serve as a good contrast to the college dining service study, as it would sample a broader, more diverse population of people.

We were pleased with how the grant application came together, and with the sense of forward momentum and commitment of a diverse interdisciplinary team representing some of the wealth of human capital that this academic and health-care community has to offer. However, when the funding announcement came, our project was not among the awardees. Without the credibility offered by the seed funding, we faced one of the persistent challenges to environmental projects: the return-on-investment dilemma. As our project planning progressed, we'd present it to various leaders—development officers, advisory boards, selected trustees—and would feel as though we had made a convincing case. Typically, people would come in skeptical, and over the course of the presentation and discussion, many would gradually "get it"—the connection we were trying to make between children, the future, health care, and the environment—and would express some enthusiasm. Some advised that we needed to hone our message and focus on distinct benefits to patient care and health rather than on prevention of environmental catastrophe or reforming the health-care industry. (Positive is better than negative.) Some said that we should focus on our need for a consolidated pediatric facility—something we had to bite our tongues to avoid pointing out that other pediatric advocates at our hospital had tried this for decades, so far without success. Only a very few who heard our presentation thought that the environmental side of the equation was compelling, and some thought it was a distraction. It just didn't feel to most of them like "our shift," seemed like an investment that would not bring in a significant return, and addressing it in the health-care context we all shared just wasn't going to provide a very compelling reward.

What about the aspect of being a "prototype"? While we were members of Practice Greenhealth and other environmental organizations, we had a long way to go to making our operating rooms less wasteful, our food healthier and more in line with environmental goals, our office recycling optimal, our energy utilization top-notch. And in doing so, might we not develop new methods that would save money, and maybe even new products that we could develop and bring to market? To make this case convincingly, we'd have to do research and have some data to show to the hospital leadership. Finding time among the hours needed for our day jobs, and having little to offer collaborators other than a sense of satisfaction, made this challenging. Like a lot of research in science, we had to prove we could get results before the experiments were done.

We started looking for other sources of funding, particularly from companies that might want to collaborate on things like energy modeling. We also started to think of creative ways to improve our operating budget. Could we build in some apartments? It will be a beautiful, nature-filled building—maybe a tower off to the side or a floor or two on top could be rental units, for physicians or administrators who might want a place to stay when on call but have a "regular" house somewhere else, for instance. We weren't sure if zoning would allow this. But these were the kinds of strategies we might need for a children's hospital. Pediatric care generally has a lower profit margin than adult care, and thus presents a more difficult business case compared to adult cancer or cardiac centers. The profit margin in a non-profit hospital is what keeps things like research, new technology, and services for patients at a superior level.

Philanthropy and Public Relations

We were starting to get the message that we were the dark horse candidate in a race for becoming one of the focus areas for the next hospital capital campaign. This is what we believed would be needed to get the all-important official commitment that this project eventually would move forward toward realization. We made the semi-final list of considerations—this was a victory—but not really as a "green hospital," but as a consolidated physical space for kids. The green idea, while appealing to the pediatric providers and families and nurses, didn't catch on with the business advisors with power at the top of the organization. Some thought it would be off-putting and wasn't compelling. From the perspective of our team, a children's hospital without this theme and mission would be just another small children's hospital, forever in the shadow of our larger relations, without a distinctive identity, and without having taken advantage of the opportunity to make a unique kind of contribution to children's health. I felt strongly that we needed a quantum leap in the identity department. My metric for success was this: we needed to do something so noteworthy, so innovative and attractive, that it would be a point-out on the Duck Boat Tours that carried people through the city every day. To achieve this vision, we needed to get a better message quickly.

Our next step was to sponsor a workshop of people from the biophilic design and architecture worlds. Julia Africa played a major role in helping to carry this off on our shoestring budget, in part by suggesting we link it temporally to a CleanMed conference that happened to be in Boston that fall. CleanMed is an annual educational and networking convention sponsored by Practice Greenhealth, which promotes environmentalism in medicine. The fact that there are conferences like this came as a surprise to me, and I suspected would to many clinicians as well. The disconnect between the passion of people attending the CleanMed convention and the unlinking of environmental concerns and health care at our organizational level provided another important lesson. We are narrow, in our niches, and often don't know what's going on down the hall, never mind in the convention center a few blocks away. We could not expect very successful and prominent business leaders who volunteer time from their packed schedules in order to serve on hospital boards to know what we were learning. John Messervy, in his quiet but persistent way, kept driving home the message that our job was to educate them.

I had never been to an architects' workshop, and so expected people to talk about engineering and blueprints and square footage. Instead, to my surprise, they talked about concepts, feelings, and mission statements. Present were people from our facilities departments and from architectural firms that had worked on our hospital network's projects. Also donating their time were national and international leaders in the world of biophilic design. One was the architect who had just completed a major children's hospital project in California, which incorporated some of the elements we wanted to include, such as design that reflected the natural world in the local region and that enveloped the patient and family experience. Another leader was Amanda Sturgeon, founder and CEO of the International Living Futures Institute (ILFI). Amanda was an architect who believed that environmental concerns were critical, that buildings had an important role to play, and that the commitment needed to go beyond just the physical structure to also include how the building worked and what values its planners and occupants espoused. This was the 360-degree approach we were hoping for, and Amanda's organization gave us a template to start thinking about what we might aspire to achieve. There were various levels of certification for buildings that actualized these principles, and the ILFI set a high bar. For instance, the building had to be more than net energy neutral—it had to actually *produce* energy. This would be a very difficult undertaking for energy-intense hospital buildings, and especially in an urban, northern climate. As John said later, it's easy for a nature education center in a cornfield in the sunbelt, but a hospital is a different matter. While other hospitals had achieved various levels of LEED certification, no hospital had ever achieved ILFI recognition as a "certified living building." Might this be the "first" to which we could aspire? I thought it was likely impossible; John seemed to relish the challenge.

While this was happening, the train was starting to roll out of the station about the focus of the capital campaign. At first, our hospital development people suggested we create a video about the project to show to prospective donors, and promptly. But then this call just faded away. It seemed decisions were being made without us, at the kinds of levels in organizations that do their work in offices that most people don't see. When I asked point blank what was going on, the development and pediatric leadership folks told me, don't worry, we can still work on finding donors, and philanthropy may be a way to move the project forward. But staff were currently preoccupied with engaging possible donors for other, more pressing priorities.

This felt to me like being told the race was over while you were still lacing your sneakers. After more than three years of methodical background work, and with chagrin at having enjoined the good will of many collaborators who had donated their time and energies in good faith to the project, I was ready to storm out and slam the door behind me. But luckily, calmer heads prevailed. In contrast to my surgical-personality impatience, with his understated manner and faintly patrician, prosodically restricted Australian diction, John just kept saying, "If I've learned one thing about this organization, it's that you just keep going until they tell you to stop. And nobody's told us to stop."

So I checked in with the pediatric departmental leadership, and got the approval to go ahead with a video. This could be used both internally and for donors. It had to be succinct and make our case more effectively than we had done for our advisory board and other decision makers. John contacted a large public relations firm that he thought would do a professional job.

It was important to me that the video be factual, that it not oversell the concept, that it reflect the research we had done, but also the mission and the passion for children's futures. That it showcased the unique opportunity with a one-of-a-kind team from a university with amazing resources and reach. That it made the case that the planet, and health, and the decisions we make in our work worlds, including health care, do matter, and are our responsibility—this too is "our shift." That this would be a prototype, and an ongoing laboratory, which might have broad impact well beyond our walls. That the time to act for the future of children must be now.

John, Hubert, and I worked on a model script and circulated it to our pediatric leadership team for edits. John found a gorgeous, artistically crafted video about a new children's hospital in Copenhagen that included some of what we had in mind. I had some patients and their families who were willing and eager to participate.

And John was right; the folks on the video production team clearly were professionals. They spent time listening to our story and read everything we sent them—the references about health and environmental effects, healing and recovery, mortality and cost. They absorbed the complex message much more quickly than our other audiences had, and synthesized our message into something brief, simple, honest, and effective. It touched on all the reasons that the reward systems of decision makers and donors might be effectively engaged in this endeavor. Agency. Altruism. Caregiving. Family bonding. Leadership. Mission. Biophilia. Doing the right thing. Being first.

The Future

As of this writing, the future of this project remains uncertain. Whether we get the big donor, convince the leadership, and bring the project to life is still a work in progress. Perhaps another place that needs a new children's hospital more than mine does will be the one to realize our goals. Perhaps we will renovate existing space using these principles and be a prototype that way, some day.

But the principles of the experiment already have been outlined. Except for a very narrow audience, changing behavior to meet environmental goals in the face of multiple competing priorities is a challenge. But even within the short few years we have been working on this project, this is changing. The hospital as a whole now has an interdisciplinary, multifaceted Center for the Environment and Health that reaches throughout the institution, including all the way to the top. We are involved with research, publication, education, advocacy, and continual tracking of metrics to improve our own operations; it's no longer seen as "weird" to make the connection between the environment and health. For many reasons, including economic and regulatory pressures and government incentives, we have made substantial progress with renewable energy and more thoughtful purchasing. A full-scale external environmental audit of our operations, including looking at all three "scopes" used by the Environmental Protection Agency to analyze an institution's footprint—on site, off site, supply chain and disposal, transportation for staff and patients, and the impact of specific investments—is underway to show objectively where making changes can have the most impact. More people have become involved and inspired by the connection between the environment and health and have volunteered to get involved or connect us with others who could be of use. We've come to recognize an untapped urge to take individual and collective action on climate change; people are worried and feel relieved when they have something they can contribute. This sense of urgency has only accelerated since we started down this path, as the effects of climate change have become more apparent even in the few years since we began. *Our brains have changed.*

And even when the environmental part of the Green Children's Hospital vision simply didn't resonate with segments of our target audience, other aspects that go along with environmentalism did. Honeybee rewards. People want good health; they want to recover; they want to have their loved ones have a good life. They want things that are beautiful and soothing and natural. They want to solve problems, face their competition, and come out ahead. They want to be rewarded for doing their jobs well and responsibly, and to have a good reputation. They want to participate in things that are meaningful.

So, while "green" didn't cut it in isolation, as predicted by our brain design, "because we should" worked less well than "because it will benefit you, here and now." B. F. Skinner, Peter Sterling, Erich Fromm, and Per Stoknes were right. Positive works better than negative. Nature attracts. Now is more convincing than later. Social rewards are powerful, diversity in talents enhances agency and fuels collaboration, and what is considered important is malleable. There is at least some hope.

Conclusion

A Sustainable Brain

WE'RE ALMOST AT THE END OF OUR EXPLORATION TOGETHER. We've taken a transcontinental walk through the history of the earth up to the present, observing the very long time during which we became what we are, and under what circumstances and for what purposes our human neural design came into being. Recall, with San Francisco marking earth's origin, multicellular organisms—the initial blueprints from which our nervous systems gradually evolved—came along in Iowa, mammals in Pennsylvania, primates in New Jersey, humans at 42nd Street in New York City, and the Anthropocene—the geological era of human-induced change to the atmosphere—only in the last 0.18 seconds of our 40-day walk. The "grand challenge" of climate change appeared long after our neural design and the human reward system evolved its complex and finely tuned inner workings, and the amazing means by which it adapted to teach us what we needed to survive and populate the earth eons before this grave new challenge could have been anticipated.

On our journey, we've stopped to peer from different perspectives at the awesome and intricate mechanism in our brain that evaluates options and makes choices. We've explored the range of human rewards, some of our inherited predispositions forged from evolutionary pressures, and our human variety and plasticity. And we've grasped the basic maxim that if something isn't perceived as rewarding to our brain, we won't do it. Before climate change was a threat, survival depended on getting more and more, and on working less to get it. Now, abruptly, our challenges have changed.

A commonly expressed "best hope" about the climate crisis is that at some point, for most people, behavior with a better environmental profile simply will become the preferred choice, without requiring a change in priorities. Top-down solutions to deliver new technologies that people will want and can afford will emerge. When alternative energy sources are cheaper and better, when carbon scrubbers can be implemented economically at scale, when electric vehicles are affordable and fun and can go the distance, people and companies will switch. The reliable transportation or the cheaper bill or the higher profit is the reward, and the environmental advantage is just the icing. When the reality of environmental disasters increasingly close to home becomes disentangled from misinformation attributing these events to non-anthropogenic causes, fear too will drive change. When people perceive and understand the cause of the problem, they will embrace new behaviors—maybe even unprecedented and potentially scary solutions like geoengineering, that likely will require entirely new kinds of cooperation among widely disparate governments.

But the evidence suggests that might take too long.

Even since our journey began, scientists from around the world in the Sixth Assessment Report by the Intergovernmental Panel on Climate Change laid out the sobering evidence that human activity–induced warming is accelerating even faster than previously predicted, increasing the intensity and duration of heat waves, droughts, fires, and storms, rupturing ecosystems and societies and raising morbidity and mortality worldwide.[1] The urgency of cutting greenhouse gas emissions has reached a "red alert" state. Before widespread invention, availability, and adoption of new technologies and large-scale institutional transformations, change has to be embraced by increasing numbers of people playing different roles and making different decisions in the present, not in some distant decade, to curb the synergistic cascades of warming that will make life on earth increasingly more difficult, dangerous, and tenuous. While the human brain is present oriented, we also have the cognitive capacity to predict future effects of our choices and already can project looking back with regret: if only more of us had done things differently sooner.

But we can draw hope from the evidence neuroscience has compiled. Humans are not fixed in stone. As we've seen in these pages, from cats to cars to drugs to sustenance, we can change, quite dramatically, in what we think is important and how we act on our shifting priorities. Our brains were designed to enable this kind of switch, using rewards to make decisions, because what is rewarding *is designed to be malleable.* This knowledge—how the brain works and adapts at the molecular, cellular, and network levels—gives us confidence that we do in fact have the capacity to change. We change at the level of the individual, and the change spreads in ways that we have come to better understand through cultures and societies. While it may not be easy and certainly is not simple, there is nothing inherent in our brain design that says that a fixed "human nature" makes us incapable of meeting this strange, urgent, novel challenge.

But to move forward, we must embrace a few hard facts.

First, don't expect change that mitigates our climate crisis to "feel" easy—or even very satisfying. This is not like other choices and changes, where you directly feel the results of your own agency, perceive the consequences of your choices tangibly, and get an immediate reward confirmed by your senses. This is not crossing the river safely or getting the Bingo match, kicking the soccer ball into the net or nailing the successful presentation at work. If CO_2 were fluorescent orange, spreading out in superterrestrial space, this would be easier. If it had a noxious smell or made our eyes water, we'd have solved this problem by now. But for this crisis, we have to rely almost exclusively on information rather than direct sensory input or how things "feel"—gut checks based on our collective past—and we have to trust unfamiliar sources for this threat that's invisible except for its muddied consequences. We can't perceive the accumulating *statistical* evidence directly with our senses. And because the effects of your thoughtfully considered, hard-won pro-environmental behavior change generally will be so distant that you won't perceive them directly, it will be difficult to "feel" any effects at all. You put in the effort to do the "right thing," even if it's inconvenient or takes resources or conflicts with another priority for which your rewards are more established, and you get . . . nothing. This characteristic of many pro environmental behavior choices, at the micro, meso, or macro level, can invite skepticism in other people, and deprives you even of some of the social rewards that can buttress many other decisions. Climate change decisions will feel inherently less satisfying than many other choices you are accustomed to making. If you understand why this is from a neural point of view, it may be less discouraging, and can help you persist.

Second, don't underestimate your influence on others. Your opinion, your decisions, your actions may have ripple effects you can't perceive. Even when you don't get immediate feedback—in fact, even when you get negative feedback—your pronouncements and priorities may influence others. People make decisions and change behavior when the sum of the millions of neural events leading up to a choice points toward change, when the reward value outweighs the risk. The knowledge of your opinions and actions transforms the weights and counterweights for other people you encounter, directly or indirectly. This is true whether your actions are pro-environmental or favor the status quo. Based on our neural design, as we've seen, people absorb what they see, compare themselves to others, and are designed to be sensitive to cultural norms.

Third, the urgency and the "wicked problem" nature of climate change and environmental compromise require that multiple solutions occur simultaneously. This challenge requires change at the individual household level, at the political level, through social movements, and through major overhauls of institutions and economic structures and incentives around the globe. Leaders and policymakers who can envision the long range, insurance experts whose projections nudge the business needle, investors looking for the next big thing, medical professionals rallying us around our own short-term health interests and concern for our families, international reporters who can connect the dots between famine and war and migration and can appeal to our altruism and the promise of hope for a better outcome, all will play a role. The lure of a less damaged environment will provide a weightier reward to more people based on these influences. Others may respond to shiny new technology that appeals, honeybee rewards, colleagues who tried new ways and liked them.

But regardless of the scale—micro, meso, macro—enough people at enough levels of decision making need to change their priorities. These levels are interdependent. The legislator will change when the voters and campaign donors change their priorities; the voters and donors will change when they perceive that the legislator aligns with their best interests. But at each level, *the individuals involved each are making decisions to change.* Each decision is made using dopamine, the prefrontal cortex evaluative machinery, the nucleus accumbens. All that we learned from disease states, from animal experiments, from imaging and single-cell recordings about how it works comes into play here. Each individual will change when the decision to change becomes more rewarding than alternative choices, and when it doesn't *feel* rewarding in the same way as you've come to expect from past decisions . . . go back and read that first hard fact!

So, with this in mind, what kinds of changes are possible? What, in a practical sense, can each of us do?

We can start with our individual choices that contribute to carbon emissions. In the sphere of our personal lives, think about the "which behaviors matter most" list from Chapter 6. What would convince you to really, truly, consistently do something different? Let's start with the example of transportation. It goes without saying that you may be boxed in by social or economic circumstances. But let's say you do have a choice. It's so convenient to drive where you need to go—but your environmental unease prompts you to give public transit a try and commute with a friend, who tells you with conviction that she wouldn't go any other way. She gets her coffee in the commuter rail station, reads on the way in, logs a brisk walk to her office so feels energized and can brag about her step count, and doesn't have the hassle or cost of driving in traffic or parking. Taking the train adds 35 minutes a day to her commute, but now she has time to finish her documentation during the return trip so she can relax once she's home. On Fridays, she treats herself to reading fiction—something she didn't have time for before. She breaks even on commuting cost, once you factor in car repair and parking, and she feels less stressed. Honeybee rewards. She's substituted one set of everyday behaviors for another. Her motivation wasn't climate change—she was forced to switch when her car was in the shop—but found it not only wasn't bad, but it had some real advantages. Substitution, not curtailment; doing differently, not doing without.

Changing your transportation may mean mass transit or getting a cleaner energy car that doesn't make your heart pump in the same way you have counted on to make decisions in the past. Moving toward a primarily plant-based diet may require that you consciously override those seductive neuronal whispers that tell you that red meat tastes *really good,* and the more the better. Those reward patterns evolved when such energy-dense food was vanishingly scarce and helped humans survive, while today, large quantities of meat really aren't in your best interest. You can figure out how to take one less air flight per year. And if you are in a position to consider family size, you can choose to factor the environment into that decision. Many in the health professions might add that you also can make the choice not to extend futile, energy-intense life-lengthening medical interventions at the end of life.[2] Having that difficult but important conversation with loved ones also can clarify what they want in this regard and can expand your ripple effect.

Even those among us who have the economic ability to make choices, care intensely about this topic, and are convinced of the science may struggle to match our own behavior to our intentions. Recall that in times of stress, people tend to lean toward the familiar. It's easier to cling to our familiar modes of "success" with their reliable formulas for recognizable rewards. This also plays out in the spheres of our choices outside our domestic lives. Are you in a leadership position in your business? Taking on the role of environmental advocacy within your company is probably outside your usual workflow, and may not help get you promoted. If you're already the boss, will this help your company grow, and your board to see you favorably, especially if the change involves an up-front financial investment, and they don't really view climate change as a problem you're being compensated to solve? Is that a risk you're willing to take? Your choices are competing with ancient neural tendencies that push us to want more tangible rewards, more than we were expecting, more than the competition. If you're a politician, is this an agenda that will help you get reelected, or is there another priority that's a better theme to gain the support of your voters? Your brain is weighing thousands of factors, with every decision, before every action. Your reward system is listening and is trying to help you. But it hasn't had a chance to catch up to the Anthropocene. It's just not predisposed to being strongly influenced by climate change.

Barring unexpected revolutionary solutions in the short term, to truly address this problem our daily decisions and customary behaviors need to change, and our individual and collective priorities need to shift, on a dizzyingly abrupt and ever-increasing scale. Based on how the neural evaluative process functions, this shift occurs from small changes in the relative weights of the factors influencing our assessment of what's rewarding from second to second. Think back to the lunch break scenario in Chapter 2; many factors were constantly shifting to influence your decision on when you packed up and left your office. Even if our most potent short-term rewards generally don't come from helping to mitigate climate change, this factor can provide the essential scale-tipping catalyst in our prefrontal evaluative decision making. Some of the counterweights have to come from sources external to us—a group-purchasing discount on energy efficient appliances; financial credits for taking the train because parking is tight. Dynamic educators at school that inspire you to take this on as a career goal—or at least as a world citizen. Tax or economic incentives for reducing waste and carbon consumption. Corporate social responsibility now becoming part of your job description and something on which your performance is judged. Community pushback on your company's new building project that makes you consider a less-polluting plan. These weights will shift with economic and political pressures from social movements, organizations, companies, voters, and governments. Decisions may be weighted, as we've seen, by input from friends and people you know, or by something you read or heard that you were predisposed to trust. Your scale may be tipped by a personal encounter with fires or floods or plastic waste, a drought-driven humanitarian crisis, a field in bloom, or responsibility for a child whose life will extend well past your own. If you pass your insights on to others, you will influence someone else's evaluative equations. Our brains integrate our direct and indirect and social experience, spritzing our decision-making pathways with tiny influences that lean our choices. These inputs can accumulate to shift what we find rewarding, and how we behave, from a small minority to a difference-making majority over time. Your changes in priority may fuel a social change, becoming part of a "tipping point" of a social epidemic within your circle of influence that will drive change on a larger scale.[3]

You can choose to be a change agent in a variety of roles you may play in your life. If social justice resonates with you, there are many people around you who have fewer choices and for whom you can advocate, as they are the most likely to be affected and have the least amount of power to adapt. You may join protests over a new dirty energy project sited in a low-income neighborhood already beset by poor air quality and high rates of asthma, or an online campaign against an overseas corporate venture that decimates the environment. You can play a change agent role at school, at work, at the community, state, or country level, and with family, friends, neighbors, co-workers, political movements, or other groups to which you belong. Information that is factual and effective is another essential contribution; some people will discover and disseminate new information through original research. If time in nature speaks to you, you can encourage others to join you, as we've seen that this can move people toward environmental action early as well as later in life. If you're able to influence children to explore wild spaces under their own agency, you may help prime future environmental advocates. If you're involved in education, your effect can be magnified.

Your ideas and examples will have the greatest effect on people you already know. Framing things positively generally works best—how you saved money with an energy efficiency intervention; how a new videoconferencing program that reduced the need for air travel had advantages for your business. If you are in a leadership position, taking this problem seriously will rub off on others around you, much more than you may realize.

Our reward system isn't a great match for reinforcing and teaching the tenuous-appearing connections between action and outcome in the realm of climate change. You likely will never see the environmental consequences of forgoing that plane trip, shifting your diet to more plant-based, making other decisions based on a carbon footprint appraisal. If you are a leader in a company, a policymaker, an educator, a healthcare CEO, a member of the board, a politician—if you chose to prioritize environmental concerns, your influence may not "pay off" in the usual way during your tenure. In fact, the environmental news is virtually guaranteed to get worse in the short term, making you and some of those who judge you question whether any of your efforts make any difference at all. It will be tempting to compare your choices to those of many others who don't factor in an environmental priority, and it will be challenging not to feel resentful if they reap short-term rewards that you're voluntarily forgoing for collective benefit. Compared to the satisfaction of crossing the river successfully, or even avoiding catching a virus, your rewards for pro-environmental choices often are limited to those you give yourself, based on information and cognition, and they're pretty weak overall. Your prefrontal evaluative processes linked to that amazing network of connections—your SNc and VTN, your hippocampus and amygdala, your nucleus accumbens and anterior cingulate cortex—will not reward you in the same way as would an ice cream sundae, a winning lottery ticket, a slam dunk, a bonus. No, you're more likely to be rewarded by reinforcement from like-minded individuals or organizations that share and reinforce your priorities by social means.

Understanding the brain's design for reward and decision making doesn't provide an instant solution, of course. But it can help. Besides illuminating our own behavioral tendencies when undertaking change in our individual lives, this knowledge may enhance our capacity to influence others and make more widespread shifts in priorities in our own sphere of influence. Collective action is crucial for the enormous "other half" of the problem, but even on the meso and macro scales, change at the level of the individual is the necessary first step to catalyze change that spreads to other people's brains and throughout organizations and governments. It can help to know, as we've noted, that you are unlikely to change people's minds—especially strangers—by heaping on more facts; people just aren't designed to respond very strongly to that approach. It can help to understand that if you are trying to influence your company, your constituency, or your town, brainstorming to identify a short-term reward that might arise from a pro-environmental decision may tip the scale toward change. As we've seen, our brains are designed to be attracted both to novelty and familiarity; keeping this in mind may help make a change more attractive. Partnering with experts in effective communication who can frame changes with legitimate positive consequences compared to business as usual may help people find rewards in new approaches to problems.

For some, revolution may resonate, involving protest and social action. Helping people to feel like part of a solution they can see—like part of a group, like someone who cares—can provide powerful social rewards. While the reward for much pro-environmental behavior at present may be weak, it is getting stronger over time. Groups like EcoTeams in Europe, Mothers Out Front in the United States, and many other advocacy and action groups show that working collectively increases your voice and magnifies mutual support and efficacy. And the social connections and sense of agency provide additional rewards—they make it genuinely positive, even fun—and give you some hope, which is why you'll stick with it. There are opportunities to write, knock on doors, demonstrate, plan events, strategize about how to persuade others, and donate time and resources. Maybe you will run for office.

And as you consider what changes you may make in your choices or how you participate in collective actions, it may be helpful to remember that in the long run, science also has shown us that *stuff doesn't make us happy.* Relationships and meaning in life are the things that correlate with long-term life satisfaction, not the fleeting rewards that were designed, by nature, for short-term survival and are evanescent.[4] When money and meaning compete, money wins the reward prize, but meaning buys happiness.

This book opened with the story of John Holter, an "ordinary person" with an extraordinary drive—saving the life of his son. While that story is legendary in my field, there are countless other examples of people making paradigm-shifting progress in a wide variety of endeavors. With enough motivation, we all have capacities that allow us to go above and beyond, to solve problems, to find solutions for reasons that matter. While our brains are better designed for efforts with immediate reward—for things that matter to us right here, right now, that we can see and touch and feel—we have talents distributed among us to do extraordinary things. Climate change is too new, and in many ways, too nebulous to align with the way we are designed to work, but with effort and persistence this crisis too can be met.

For certain, meeting this grand challenge will be difficult. Now is when we must alter, using our cognition, the weight of what is rewarding, while our brains do their best to adapt to accelerating change, and science keeps moving forward, inventing better tools to satisfy our insatiable demand for energy. And for stuff—although we can work some on that tendency. This challenge will take discipline and persistence and going out on a limb, and despite eco-hedonism and honeybee approaches and the "better quality of life" claims, there's simply no way this won't mean changes that we will make kicking and screaming. We will have to give up habits and make some difficult choices and sacrifices; there are conveniences and basic things we will have to do without. We will need to keep in mind how much worse this is for people who are not in the 20-tons-of-carbon-emissions-a-year class. Undoubtedly the conflicts we already see on many scales will increase—micro, meso, macro; people, businesses, governments, countries—and there will be consequences that will be painful and destructive. There can be no doubt that there will be much suffering, and loss, and mourning; still, we remain with some control over the scale of these tragedies, based on our actions now.

But since we have learned that positive is better than negative, that a sense of agency is critical to motivation for action, that social supports can flip the switch, that we do have the capacity to change—we have the means to move forward. Each individual brings a unique collection of biological traits shaped by experience, with diverse talents that evolution designed. Some will care about nature, some about equity, some about organizations, some about business advantages, some about government, some about science, and some about innovation and discovery. Likely there will be some unexpected alliances.

Meeting this nothing-like-it-ever-before grand challenge will take every bit of the capacity of the most miraculous invention ever made by evolution. We have only just begun the exploration of how the brain and climate change intersect, but we know enough now to choose our next steps. Neuroscience reveals our inherent ability to change ourselves and those whose lives we touch, to make choices that will sustain us, and to pass on to others what we have learned to value and prioritize today. Our brains got us here, and our sustainable brains are the only hope to get us to a better future.

NOTES

ACKNOWLEDGMENTS

INDEX

Notes

PREFACE

1. McKibben B. A special moment in history. The Atlantic; May 1998. [Available at https://www.theatlantic.com/magazine/archive/1998/05/a-special-moment-in-history/377106/.]

INTRODUCTION

1. Rachel RA. Surgical treatment of hydrocephalus: a historical perspective. Pediatric Neurosurgery. 1999;30:296–304; Hayward R. "Casey and Theo": the children who changed the face of "water-on-the-brain." British Journal of Neurosurgery. 2009; 23(4):347–50.
2. Dietz T, Gardner GT, Gilligan J, et al. Household actions can provide a behavioral wedge to rapidly reduce US carbon emissions. Proceedings of the National Academy of Sciences USA. 2009;106(44):18452; Pacala S, Socolow R. Stabilization wedges: solving the climate problem for the next 50 years with current technologies. Science. 2004(305): 968–72; Schnoor JL. Coalitions of the willing. Environmental science & technology. 2012;46(17):9201; Girod B, van Vuuren DP, Hertwich EG. Climate policy through changing consumption choices: options and obstacles for reducing greenhouse gas emissions. Global Environmental Change. 2014;25(March):5-3780; Wynes S, Nicholas KA. The climate mitigation gap: education and government recommendations miss the most effective individual actions. Environmental Research Letters. 2017;12(7):074024.
3. McKibben B. A special moment in history, part 2: Earth 2. The Atlantic 1998;281(5):55–78; Intergovernmental Panel on Climate Change. IPCC 2018: summary for policymakers. In: Masson-Delmotte V, Zhai P, Pörtner HO, et al., editors. Global Warming of 15°C. An IPCC Special Report on the impacts of global warming of 15°C above pre-industrial levels and related global greenhouse gas emission pathways, in the context of strengthening the global response to the threat of climate change, sustainable

development, and efforts to eradicate poverty. Geneva: World Meteorologic Organization; 2018, pp. 1–32; Nature Editorial Board. Governments must take heed of latest IPCC assessment. (Intergovernmental Panel on Climate Change) (Report) Nature 2018;562(7726):163; Diaz S, Settele J, Brondizio E, et al. Summary for policymakers of the global assessment report on biodiversity and ecosystem services of the Intergovernmental Science-Policy Platform on Biodiversity and Ecosystem Services. Intergovernmental Science-Platform on Biodiversity and Ecosystem Services 2019; Henriques ST, Borowiecki KJ. The drivers of long-run CO_2 emissions in Europe, North America and Japan since 1800. Energy Policy. 2017;101:537–49.

4. Swim J, Clayton S, Doherty T, et al. Psychology and global climate change: addressing a multi-faceted phenomenon and set of challenges. A report by the American Psychological Association's Task Force on the Interface between Psychology and Global Climate Change. American Psychological Association, 2011 [Available from: http://www.apa.org/science/about/publications/climate-change.aspx.]; Kahneman D. Thinking, fast and slow. First edition. New York: Farrar, Straus and Giroux, 2011.
5. McQueen A, Cress C, Tothy A. Using a tablet computer during pediatric procedures: a case series and review of the "apps." Pediatric Emergency Care 2012;28(7):712–14.
6. Schrag DP. Geobiology of the Anthropocene. In: Knoll AH, Canfield DE, Konhauser KO, editors. Fundamentals of Geobiology. Oxford: Blackwell Publishing; 2012, pp. 425–36.
7. United Nations. UN Report: Nature's dangerous decline "unprecedented"; species extinction rates "accelerating" 2019 [Available from: https://www.un.org/sustainabledevelopment/blog/2019/05/nature-decline-unprecedented-report/]; Diaz S, Settele J, Brondizio E, et al. Summary for policymakers of the global assessment report on biodiversity and ecosystem services of the Intergovernmental Science-Policy Platform on Biodiversity and Ecosystem Services. Intergovernmental Science-Platform on Biodiversity and Ecosystem Services; Bonn, Germany: IPBES Secretariat, 2019.
8. NATO, National security and human health implications of climate change. Fernando HJS, Klaic ZB, McCulley JL, editors. Dordrecht: Springer; 2012; Intergovernmental Panel on Climate Change. IPCC 2018: summary for policymakers. In: Masson-Delmotte V, Zhai P, Pörtner HO, et al., editors. Global warming of 15°C: an IPCC special report on the impacts of global warming of 15°C above pre-industrial levels and related global greenhouse gas emission pathways, in the context of strengthening the global response to the threat of climate change, sustainable development, and efforts to eradicate poverty. Geneva: World Meteorologic Organization; 2018, pp. 1–32.

1. BRAIN EVOLUTION AND THE ANTHROPOCENE

1. Creely H, Khaitovich P. Human brain evolution. Progress in Brain Research. 2006;158: 295–309; Martin RD. Human brain evolution in an ecological context. 52nd James Arthur lecture on the evolution of the human brain. New York: American Museum of Natural History; 1982; de Sousa A, Cunha E. Hominins and the emergence of the modern human brain. Progress in Brain Research. 2012;195:293–322.

2. Schrag DP. Geobiology of the Anthropocene. In: Knoll AH, Canfield DE, Konhauser KO, editors. Fundamentals of geobiology: Blackwell Publishing; 2012, pp. 425–36; Crutzen PJ. Geology of mankind. Nature. 2002;415(6867):23; Crutzen P, Steffen W. How long have we been in the Anthropocene Era? Climatic Change. 2003;61(3):251–57; Waters CN, Zalasiewicz J, Summerhayes C, et al. The Anthropocene is functionally and stratigraphically distinct from the Holocene. Science. 2016;351(6269):137–37.

3. Sterling P, Laughlin S. Principles of neural design. Cambridge, MA: MIT Press; 2015; Zhao K, Liu M, Burgess R. Adaptation in bacterial flagellar and motility systems: from regulon members to 'foraging'-like behavior in *E-coli*. Nucleic Acids Research. 2007; 35(13):4441–52.

4. Yang Y, Sourjik V. Opposite responses by different chemoreceptors set a tunable preference point in *Escherichia coli* pH taxis. Molecular Microbiology. 2012;86(6):1482–89; Zhao K, Liu M, Burgess R. Adaptation in bacterial flagellar and motility systems: from regulon members to 'foraging'-like behavior in *E-coli*. Nucleic Acids Research. 2007;35(13):4441–52.

5. Hekimi S. A neuron-specific antigen in *C. elegans* allows visualization of the entire nervous system. Neuron. 1990;4(6):855–65; Haspel G, O'Donovan MJ. A connectivity model for the locomotor network of *Caenorhabditis elegans*. Worm. 2012;1(2):125.

6. Baxter DA, Byrne JH. Feeding behavior of *Aplysia:* a model system for comparing cellular mechanisms of classical and operant conditioning. Learning & Memory. 2006;13(6):669–80.

7. Hebb DO. The organization of behavior: a neuropsychological theory. New York: Wiley; 1949; Löwel S, Singer W. Selection of intrinsic horizontal connections in the visual cortex by correlated neuronal activity. Science. 1992;255(5041):209–12.

8. Huber R, Panksepp JB, Nathaniel T, Alcaro A, Panksepp J. Drug-sensitive reward in crayfish: an invertebrate model system for the study of SEEKING, reward, addiction, and withdrawal. Neuroscience and Biobehavioral Reviews. 2011;35(9):1847–53; Brembs B. Spontaneous decisions and operant conditioning in fruit flies. Behavioural Processes. 2011;87(1):157–64; Waddell S. Reinforcement signalling in *Drosophila;* dopamine does it all after all. Current Opinion in Neurobiology. 2013;23(3):324–29; Hawkins RD, Byrne JH. Associative learning in invertebrates. Cold Spring Harbor Perspectives in Biology. 2015;7(5):1–18.

9. Changizi MA, McGehee RMF, Hall WG. Evidence that appetitive responses for dehydration and food-deprivation are learned. Physiology & Behavior. 2002;75(3):295–304.

10. Darwin C. On the origin of species: by means of natural selection, or, the preservation of favored races in the struggle for life. London: John Murray; 1859.

11. Schultz W. Neuronal reward and decision signals: from theories to data. Physiological Reviews. 2015;95(3):853–951.

2. BRAIN REWARDS AS A DESIGN FOR LEARNING

1. Herculano-Houzel S. The human brain in numbers: a linearly scaled-up primate brain. Frontiers of Human Neuroscience. 2009;3:31; Azevedo FA, Carvalho LR, Grinberg LT, et al. Equal numbers of neuronal and nonneuronal cells make the human brain an isometrically scaled-up primate brain. Journal of Comparative Neurology. 2009;513(5): 532–41; Pakkenberg B, Pelvig D, Marner L, et al. Aging and the human neocortex. Experimental Gerontology. 2003;38:95–9.
2. Sterling P, Laughlin S. Principles of neural design. Cambridge, MA: MIT Press; 2015.
3. Hawkins J, Ahmad S. Why neurons have thousands of synapses, a theory of sequence memory in neocortex. Front Neural Circuits. 2016;10:23.
4. Harlow JM. Passage of an iron rod through the head. Boston Medical and Surgical Journal. 1848;39(20):389–93; Barker FG. Phineas among the phrenologists: the American crowbar case and nineteenth-century theories of cerebral localization. Journal of Neurosurgery. 1995;82:672–82.
5. Carlsson A. Nobel lecture. A half-century of neurotransmitter research: impact on neurology and psychiatry. Nobelprize.org; 2000. [Available from: http://www.nobelprize.org/nobel_prizes/medicine/laureates/2000/carlsson-lecture.html.]
6. Goldman JG, Goetz CG. History of Parkinson's disease. Handbook of Clinical Neurology. 2007;83:109–128; Sacks O. Awakenings. Revised ed. Harmondsworth, England: Penguin Books; 1976; Sacks O. The origin of "awakenings." BMJ (Clinical Research Edition). 1983;287(6409):1968; Gale JT, Amirnovin R, Williams ZM, et al. From symphony to cacophony: pathophysiology of the human basal ganglia in Parkinson disease. Neuroscience and Biobehavioral Reviews. 2008;32(3):378–87.
7. Bonvin C, Horvath J, Christe B, et al. Compulsive singing: another aspect of punding in Parkinson's disease. Annals of Neurology. 2007;62(5):525–28.
8. Weintraub D, Koester J, Potenza MN, et al. Impulse control disorders in Parkinson disease: a cross-sectional study of 3090 patients. Archives of Neurology. 2010; 67(5):589.
9. Aleksandrova LR, Creed MC, Fletcher PJ, et al. Deep brain stimulation of the subthalamic nucleus increases premature responding in a rat gambling task. (Report.) Behavioural Brain Research. 2013;245:76.
10. Aleksandrova, Creed, Fletcher, et al. Deep brain stimulation of the subthalamic nucleus.
11. Piano AN, Tan LCS. Impulse control disorder in a patient with X-linked dystonia-Parkinsonism after bilateral pallidal deep brain stimulation. Parkinsonism & Related Disorders. 2013;19:1069–70; Witjas T, Baunez C, Henry JM, et al. Addiction in Parkinson's disease: impact of subthalamic nucleus deep brain stimulation. Movement Disorders. 2005;20(8):1052–55; Stefani A, Galati S, Brusa L, et al. Pathological gambling from dopamine agonist and deep brain stimulation of the nucleus tegmenti pedunculopontine. BMJ Case Reports. 2010;2010:1–3.

12. Patel SR, Sheth SA, Mian MK, et al. Single-neuron responses in the human nucleus accumbens during a financial decision-making task. Journal of Neuroscience. 2012;32(21):7311.
13. Stenstrom E, Saad G. Testosterone, financial risk-taking, and pathological gambling. Journal of Neuroscience, Psychology, and Economics. 2011;4(4):254–66.
14. Bechara A, Damasio AR, Damasio H, Anderson SW. Insensitivity to future consequences following damage to human prefrontal cortex. Cognition. 1994;50:7–15; Szczepanski SM, Knight RT. Insights into human behavior from lesions to the prefrontal cortex. Neuron. 2014;83(5):1002–18; Evens R, Stankevich Y, Dshemuchadse M, et al. The impact of Parkinson's disease and subthalamic deep brain stimulation on reward processing. Neuropsychologia. 2015;75:11–19.
15. Ballard IC, Murty VP, Carter RM, et al. Dorsolateral prefrontal cortex drives mesolimbic dopaminergic regions to initiate motivated behavior. Journal of Neuroscience. 2011;31(28):10340; Bahlmann J, Aarts E, D'Esposito M. Influence of motivation on control hierarchy in the human frontal cortex. Journal of Neuroscience. 2015;35(7):3207; Diekhof EK, Gruber O. When desire collides with reason: functional interactions between anteroventral prefrontal cortex and nucleus accumbens underlie the human ability to resist impulsive desires. Journal of Neuroscience. 2010;30(4):1488; Labudda K, Brand M, Mertens M, et al. Decision making under risk condition in patients with Parkinson's disease: a behavioural and fMRI study. Behavioural Neurology. 2010;23(3): 131–43; Kühn S, Romanowski A, Schilling C, et al. The neural basis of video gaming. Translational Psychiatry. 2011:e53.
16. Lataster J, Collip D, Ceccarini J, et al. Psychosocial stress is associated with in vivo dopamine release in human ventromedial prefrontal cortex: a positron emission tomography study using [18F]fallypride. NeuroImage. 2011;58(4):1081–89; Tian M, Chen Q, Zhang Y, et al. PET imaging reveals brain functional changes in internet gaming disorder. European Journal of Nuclear Medicine & Molecular Imaging 2014;41(7):1388–97; Egerton A, Mehta MA, Montgomery AJ, et al. The dopaminergic basis of human behaviors: a review of molecular imaging studies. Neuroscience and Biobehavioral Reviews. 2009;33(7):1109–32.
17. Kassubek J, Abler B, Pinkhardt EH. Neural reward processing under dopamine agonists: Imaging. Journal of the Neurological Sciences. 2011;310(1–2):36–39; Atlas LY, Whittington RA, Lindquist MA, et al. Dissociable influences of opiates and expectations on pain. Journal of Neuroscience. 2012;32(23):8053–64; Brattico E, Jacobsen T, Vartiainen N, et al. A functional MRI study of happy and sad emotions in music with and without lyrics. Frontiers in Psychology. 2011;2:1–16; Hagerty MR, Isaacs J, Brasington L, et al. Case study of ecstatic meditation: fMRI and EEG evidence of self-stimulating a reward system. Journal of Neural Transplantation & Plasticity. 2013;2013:653572–12; Bruneau EG, Jacoby N, Saxe R. Empathic control through coordinated interaction of amygdala, theory of mind and extended pain matrix brain regions. Neuroimage. 2015;114:105–19; Falk EB, Way BM, Jasinska AJ. An imaging genetics approach to understanding social influence. Frontiers in Human Neuroscience.

2012;6:168(1–13); Chiew KS, Braver TS. Positive affect versus reward: emotional and motivational influences on cognitive control. Frontiers in Psychology. 2011;2:279; Cox SML, Andrade A, Johnsrude IS. Learning to like: a role for human orbitofrontal cortex in conditioned reward. Journal of Neuroscience. 2005;25(10):2733; Leknes S, Lee M, Berna C, et al. Relief as a reward: hedonic and neural responses to safety from pain. PloS One. 2011;6(4):e17870; Panksepp J. Feeling the pain of social loss. Science. 2003;302(5643):237–39; Sailer U, Robinson S, Fischmeister F, et al. Altered reward processing in the nucleus accumbens and mesial prefrontal cortex of patients with posttraumatic stress disorder. Neuropsychologia. 2008;46(11):2836–44; Macoveanu J. Serotonergic modulation of reward and punishment: evidence from pharmacological fMRI studies. Brain Research. 2014;1556:19–27; Swain J, Kim P, Spicer J, et al. Approaching the biology of human parental attachment: brain imaging, oxytocin and coordinated assessments of mothers and fathers. Brain Research. 2014;1580:78–101; Geiger N, Bowman C, Clouthier T, et al. Observing environmental destruction stimulates neural activation in networks associated with empathic responses. Social Justice Research. 2017;30(4):300–22.

18. Dreher JC, Meyer-Lindenberg A, Kohn P, Berman KF. Age-related changes in midbrain dopaminergic regulation of the human reward system. Proceedings of the National Academy of Sciences of the United States of America. 2008;105(39):15106–11.

19. Wise RA. Dopamine, learning and motivation. Nature Reviews Neuroscience. 2004;5(6):483; Asaad WF, Eskandar EN. Encoding of both positive and negative reward prediction errors by neurons of the primate lateral prefrontal cortex and caudate nucleus. Journal of Neuroscience. 2011;31(49):17772.

20. Schultz W, Dayan P, Montague PR. A neural substrate of prediction and reward. Science. 1997;275(5306):1593–99; Schultz W. Behavioral theories and the neurophysiology of reward. Annual Review of Psychology. 2006;57:87–115.

21. Schultz W. Multiple dopamine functions at different time courses. Annual Review of Neuroscience. 2007;30:259–88.

22. Schultz W. Neuronal reward and decision signals: from theories to data. Physiological Reviews. 2015;95(3):853–951.

23. Masayuki M, Okihide H. Two types of dopamine neuron distinctly convey positive and negative motivational signals. Nature. 2009;459(7248):837.

24. Owesson-White CA, Cheer JF, Beyene M, et al. Dynamic changes in accumbens dopamine correlate with learning during intracranial self-stimulation. Proceedings of the National Academy of Sciences of the United States of America. 2008;105(33): 11957; Schluter E, Mitz A, Cheer J, Averbeck B. Real-time dopamine measurement in awake monkeys. PloS One. 2014;9(6); Howard CD, Daberkow DP, Ramsson ES, et al. Methamphetamine-induced neurotoxicity disrupts naturally occurring phasic dopamine signaling. European Journal of Neuroscience. 2013;38(1):2078–88; Willuhn I, Tose A, Wanat MJ, et al. Phasic dopamine release in the nucleus accumbens in response to pro-social 50 kHz ultrasonic vocalizations in rats. Journal

of Neuroscience. 2014;34(32):10616; Nakazato T. Dual modes of extracellular serotonin changes in the rat ventral striatum modulate adaptation to a social stress environment, studied with wireless voltammetry. Experimental Brain Research. 2013;230(4):583–96.

25. Robinson DL, Heien MLAV, Wightman RM. Frequency of dopamine concentration transients increases in dorsal and ventral striatum of male rats during introduction of conspecifics. Journal of Neuroscience. 2002;22(23):10477.

26. Kishida KT, Saez I, Lohrenz T, et al. Subsecond dopamine fluctuations in human striatum encode superposed error signals about actual and counterfactual reward. Proceedings of the National Academy of Sciences of the United States of America. 2016;113(1):200.

27. Sterling P. Why we consume: neural design and sustainability: Great Transition Initiative; 2016. [Available from: http://www.greattransition.org/publication/why-we-consume]; Rees WE. Human nature, eco-footprints and environmental injustice. Local Environment. 2008;13(8):685–701.

28. Trotzke P, Starcke K, Muller A, Brand M. Pathological buying online as a specific form of internet addiction: a model-based experimental investigation. (Report). PLOS One. 2015;10(10); Lawrence LM, Ciorciari J, Kyrios M. Relationships that compulsive buying has with addiction, obsessive-compulsiveness, hoarding, and depression. Comprehensive Psychiatry. 2014;55(5):1137–45.

29. West R. Theories of addiction. Addiction. 2001;96:3–13; Heyman GM. Addiction and choice: theory and new data. Frontiers of Psychiatry. 2013;4:31.

30. Huber R, Panksepp JB, Nathaniel T, et al. Drug-sensitive reward in crayfish: An invertebrate model system for the study of SEEKING, reward, addiction, and withdrawal. Neuroscience and Biobehavioral Reviews. 2011;35(9):1847–53; Sovik E, Barron AB. Invertebrate Models in Addiction Research. Brain, Behavior and Evolution. 2013;82(3):153–65.

31. Wink M, Schmeller T, Latz-Brüning B. Modes of action of allelochemical alkaloids: interaction with neuroreceptors, DNA, and other molecular targets. Journal of Chemical Ecology. 1998;24(11):1881–937; Berridge KC, Kringelbach ML. Affective neuroscience of pleasure: reward in humans and animals. Psychopharmacology (Berl). 2008;199(3):457–80; Robison AJ, Nestler EJ. Transcriptional and epigenetic mechanisms of addiction. Nature Reviews Neuroscience. 2011;12(11):623.

32. Redish AD. Addiction as a computational process gone awry. Science. 2004;306(5703):1944–47.

33. Belin D, Belin-Rauscent A, Murray JE, Everitt BJ. Addiction: failure of control over maladaptive incentive habits. Current Opinion in Neurobiology. 2013;23(4):564–72; Clemens KJ, Castino MR, Cornish JL, et al. Behavioral and neural substrates of habit formation in rats intravenously self-administering nicotine. Neuropsychopharmacology. 2014;39(11):2584.

34. Smith KS, Graybiel AM. Habit formation coincides with shifts in reinforcement representations in the sensorimotor striatum. Journal of Neurophysiology. 2016;115(3):1487; Schwabe L, Wolf OT. Stress prompts habit behavior in humans. Journal of Neuroscience. 2009;29(22):7191; Barker JM, Taylor JR, Chandler LJ. A unifying model of the role of the infralimbic cortex in extinction and habits. Learning & Memory. 2014;21(9):441–48.

35. Balleine BW, Dezfouli A. Hierarchical action control: adaptive collaboration between actions and habits. Frontiers in Psychology. 2019;10:2735; Smith, Graybiel. Habit formation coincides with shifts in reinforcement representations; Yin HH, Knowlton BJ. The role of the basal ganglia in habit formation. Nature Reviews Neuroscience 2006;7(6):464–76.

36. West R. Theories of addiction. Addiction. 2001;96:3–13; Robinson TE, Berridge KC. Addiction. Annual Review of Psychology. 2003:25; Robinson TE, Berridge KC. Review. The incentive sensitization theory of addiction: some current issues. Philosophical Transactions of the Royal Society of London Series B, Biological Sciences. 2008; 363(1507):3137; Koob GF, Le Moal M. Review. Neurobiological mechanisms for opponent motivational processes in addiction. Philosophical Transactions of the Royal Society of London Series B, Biological Sciences. 2008;363(1507):3113–23; Belin, Belin-Rauscent, Murray, Everitt. Addiction: failure of control.

37. Ross DR, Finestone DH, Lavin GK. Space Invaders obsession. JAMA. 1982;248(10): 1177; Harry B. Obsessive video-game users. JAMA. 1983;249(4):473; Keepers GA. Pathological preoccupation with video games. Journal of the American Academy of Child & Adolescent Psychiatry. 1990;29(1):49–50; Mitchell P. Internet addiction: genuine diagnosis or not? The Lancet. 2000;355(9204):632; Hellman M, Schoenmakers TM, Nordstrom BR, Van Holst RJ. Is there such a thing as online video game addiction? A cross-disciplinary review. Addiction Research & Theory. 2013;21(2): 102–12; Meerkerk GJ, Van Den Eijnden RJJM, Vermulst AA, Garretsen HFL. The Compulsive Internet Use Scale (CIUS): some psychometric properties. Cyberpsychology & Behavior. 2009;12(1):1; Van Rooij AJ, Schoenmakers TM, Vermulst AA, et al. Online video game addiction: identification of addicted adolescent gamers. Addiction (Abingdon, England). 2011;106(1):205; Jelenchick LA, Eickhoff J, Christakis DA, et al. The problematic and risky internet use screening scale (PRIUSS) for adolescents and young adults: Scale development and refinement. Computers in Human Behavior. 2014;35:171–78; Cho H, Kwon M, Choi J-H, et al. Development of the Internet addiction scale based on the Internet Gaming Disorder criteria suggested in DSM-5. Addictive Behaviors. 2014;39(9):1361–66; Loredana V, Gioacchino L, Santo DIN. Buying addiction: reliability and construct validity of an assessment questionnaire. Postmodern Openings. 2015;VI(1):149–60.

38. Weinstein A, Weizman A. Emerging association between addictive gaming and attention-deficit / hyperactivity disorder. Current Psychiatry Reports. 2012;14(5):590–97; Kardefelt-Winther D. A conceptual and methodological critique of internet addiction research: towards a model of compensatory internet use. Computers in Human

Behavior. 2014;31:351; Wu K, Politis M, O' Sullivan SS, et al. Problematic internet use in Parkinson's disease. Parkinsonism and Related Disorders. 2014;20(5):482–87.

39. Koepp MJ, Gunn RN, Lawrence AD, et al. Evidence for striatal dopamine release during a video game. Nature. 1998;393(6682):266; Tian, Chen, Zhang, et al. PET imaging reveals brain functional changes; Kühn, Romanowski, Schilling, et al. The neural basis of video gaming; Mathiak KA, Klasen M, Weber R, et al. Reward system and temporal pole contributions to affective evaluation during a first person shooter video game. (Report). BMC Neuroscience. 2011;12:66; Lorenz RC, Gleich T, Gallinat J, Kuhn S. Video game training and the reward system. Frontiers in Human Neuroscience. 2015;9:40(1–9); Han DH, Bolo N, Daniels MA, et al. Brain activity and desire for internet video game play. Comprehensive Psychiatry. 2011; 52(1):88–95.

40. Pignatelli M, Bonci A. Role of dopamine neurons in reward and aversion: a synaptic plasticity perspective. Neuron. 2015;86(5):1145.

41. Sabatinelli D, Bradley MM, Lang PJ, et al. Pleasure rather than salience activates human nucleus accumbens and medial prefrontal cortex. Journal of Neurophysiology. 2007;98(3):1374.

42. McGann J. Poor human olfaction is a 19th-century myth. Science. 2017;356(6338):1–6.

43. Olry R, Haines D. NEUROwords: From Dante Alighieri's First Circle to Paul Donald MacLean's Limbic System. Journal of the History of the Neurosciences. 2005;14(4):368–70; Catani M, Dell'Acqua F, De Schotten MT. A revised limbic system model for memory, emotion and behaviour. Neuroscience and Biobehavioral Reviews. 2013;37(8):1724; Berridge, Kringelbach. Affective neuroscience of pleasure; Montague PR, Hyman SE, Cohen JD. Computational roles for dopamine in behavioural control. Nature. 2004; 431:760–67.

44. Sterling, Laughlin. Principles of neural design.

45. Adachi M, Hosoya T, Haku T, et al. Evaluation of the substantia nigra in patients with Parkinsonian syndrome accomplished using multishot diffusion-weighted MR imaging. American Journal of Neuroradiology. 1999;20(8):1500; D'Ardenne K, McClure SM, Nystrom LE, Cohen JD. BOLD responses reflecting dopaminergic signals in the human ventral tegmental area. Science. 2008;319(5867):1264.

46. Wise. Dopamine, learning and motivation; Arias-Carrion O, Stamelou M, Murillo-Rodriguez E, et al. Dopaminergic reward system: a short integrative review. International Archives of Medicine. 2010;3(24):1–6.

47. Björklund A, Dunnett SB. Dopamine neuron systems in the brain: an update. Trends in Neurosciences. 2007;30(5):194–202.

48. Berridge, Kringelbach. Affective neuroscience of pleasure.

49. Schultz. Behavioral theories and the neurophysiology; Schultz. Multiple dopamine functions; Nomoto K, Schultz W, Watanabe T, Sakagami M. Temporally extended dopamine responses to perceptually demanding reward-predictive stimuli. Journal of

Neuroscience. 2010;30(32):10692–702; Schultz. Neuronal reward and decision signals; Schultz W. Dopamine reward prediction-error signalling: a two-component response. Nature Reviews Neuroscience 2016;17(3):183–95; Montague, Hyman, Cohen. Computational roles for dopamine.

50. Schultz. Neuronal reward and decision signals.

51. Berridge, Kringelbach. Affective neuroscience of pleasure; Catani, Dell'Acqua, De Schotten. A revised limbic system model; Chang SWC, Fagan NA, Toda K, et al. Neural mechanisms of social decision-making in the primate amygdala. Proceedings of the National Academy of Sciences of the United States of America. 2015;112(52): 16012; Ahn S, Phillips AG. Modulation by central and basolateral amygdalar nuclei of dopaminergic correlates of feeding to satiety in the rat nucleus accumbens and medial prefrontal cortex. Journal of Neuroscience. 2002;22(24):10958–65.

52. Koscik TR, Tranel D. The human amygdala is necessary for developing and expressing normal interpersonal trust. Neuropsychologia. 2011;49(4):602–11; Bruneau, Jacoby, Saxe. Empathic control through coordinated interaction; Ousdal OT, Specht K, Server A, et al. The human amygdala encodes value and space during decision making. Neuroimage. 2014;101:712.

53. Bruneau, Jacoby, Saxe. Empathic control through coordinated interaction; Burgos-Robles A, Kimchi EY, Izadmehr EM, et al. Amygdala inputs to prefrontal cortex guide behavior amid conflicting cues of reward and punishment. Nature Neuroscience. 2017;20(6):824–835; Prévost C, McCabe JA, Jessup RK, et al. Differentiable contributions of human amygdalar subregions in the computations underlying reward and avoidance learning. European Journal of Neuroscience. 2011;34(1):134; Schlund MW, Cataldo MF. Amygdala involvement in human avoidance, escape and approach behavior. Neuroimage. 2010;53(2):769–76; Izquierdo A, Darling C, Manos N, et al. Basolateral amygdala lesions facilitate reward choices after negative feedback in rats. Journal of Neuroscience. 2013;33(9):4105; Li J, Schiller D, Schoenbaum G, et al. Differential roles of human striatum and amygdala in associative learning. Nature Neuroscience. 2011;14(10):1250–52.

54. Bahlmann, Aarts, D'Esposito. Influence of motivation on control hierarchy.

55. Botvinick M, Braver T. Motivation and cognitive control: from behavior to neural mechanism. Annual Review of Psychology. 2015;66:83–113.

56. Seo H, Lee D. Temporal filtering of reward signals in the dorsal anterior cingulate cortex during a mixed-strategy game. Journal of Neuroscience. 2007;27(31):8366–77.

57. Schultz, Dayan, Montague. A neural substrate of prediction and reward; Montague, Hyman, Cohen. Computational roles for dopamine.

58. Fladung A-K, Gron G, Grammer K, et al. A neural signature of anorexia nervosa in the ventral striatal reward system. American Journal of Psychiatry. 2010;167(2):206.

59. Tremblay L, Schultz W. Relative reward preference in primate orbitofrontal cortex. Nature. 1999;398(6729):704–8.

60. Tremblay, Schultz. Relative reward preference; Roesch MR, Olson CR. Neuronal activity related to reward value and motivation in primate frontal cortex. Science. 2004;304(5668):307–10.

61. Padoa-Schioppa C, Assad JA. Neurons in the orbitofrontal cortex encode economic value. Nature. 2006;441(7090):223–26.

62. Howe MW, Tierney PL, Sandberg SG, et al. Prolonged dopamine signalling in striatum signals proximity and value of distant rewards. (Report). Nature. 2013; 500(7464):575; Hare TA, Camerer CF, Rangel A. Self-control in decision-making involves modulation of the vmPFC valuation system. Science. 2009;324(5927)):646–48.

63. Epstein S. Integration of the cognitive and the psychodynamic unconscious. American Psychologist. 1994;49(8):709–24; Sloman SA. The empirical case for two systems of reasoning. Psychological Bulletin. 1996;119(1):3–22; Hofmann W, Friese M, Strack F. Impulse and self-control from a dual-systems perspective. Perspectives on Psychological Science. 2009;4(02):162–76; Kahneman D. Thinking, fast and slow. 1st ed. New York: Farrar, Straus and Giroux; 2011.

64. Kandel ER. The molecular biology of memory storage: A dialogue between genes and synapses. Science. 2001;294(5544):1030–38; Dawkins R. The selfish gene. 30th anniversary ed. Oxford: Oxford University Press; 2006.

65. Knight EJ, Min H-K, Hwang S-C, et al. Nucleus accumbens deep brain stimulation results in insula and prefrontal activation: a large animal fMRI study. PLoS One. 2013;8(2); Richardson NR, Gratton A. Changes in medial prefrontal cortical dopamine levels associated with response-contingent food reward: an electrochemical study in rat. Journal of Neuroscience. 1998;18(21):9130; Wise RA. Role of brain dopamine in food reward and reinforcement. Philosophical Transactions: Biological Sciences. 2006;361(1471):1149–58; Montague, Hyman, Cohen. Computational roles for dopamine.

66. Cone JJ, Roitman JD, Roitman MF. Ghrelin regulates phasic dopamine and nucleus accumbens signaling evoked by food-predictive stimuli. Journal of Neurochemistry. 2015;133(6):844–56; Krügel U, Schraft T, Kittner H, et al. Basal and feeding-evoked dopamine release in the rat nucleus accumbens is depressed by leptin. European Journal of Pharmacology. 2003;482(1):185–87; Aitken TJ, Greenfield VY, Wassum KM. Nucleus accumbens core dopamine signaling tracks the need-based motivational value of food-paired cues. Journal of Neurochemistry. 2016;136(5):1026–36; Calipari ES, Espana RA. Hypocretin / orexin regulation of dopamine signaling: implications for reward and reinforcement mechanisms. Frontiers of Behavioural Neuroscience. 2012;(1–13):54; Changizi MA, McGehee RMF, Hall WG. Evidence that appetitive responses for dehydration and food-deprivation are learned. Physiology & Behavior. 2002;75(3):295–304.

67. Luo AH, Tahsili-Fahadan P, Wise RA, et al. Linking context with reward: a functional circuit from hippocampal CA3 to ventral tegmental area. (Report). Science. 2011;333 (6040):353.

68. McCutcheon JE. The role of dopamine in the pursuit of nutritional value. Physiology & Behavior. 2015;152:408–15.
69. Aitken, Greenfield, Wassum. Nucleus accumbens core dopamine signaling; Cone JJ, Fortin SM, McHenry JA, et al. Physiological state gates acquisition and expression of mesolimbic reward prediction signals. Proceedings of the National Academy of Sciences of the United States of America. 2016;113(7):1943.
70. Brown HD, McCutcheon JE, Cone JJ, et al. Primary food reward and reward-predictive stimuli evoke different patterns of phasic dopamine signaling throughout the striatum. European Journal of Neuroscience. 2011;34(12):1997–2006; Nakazato T. Striatal dopamine release in the rat during a cued lever-press task for food reward and the development of changes over time measured using high-speed voltammetry. Experimental Brain Research. 2005;166(1):137–46; Yoshimi K, Kumada S, Weitemier A, et al. Reward-induced phasic dopamine release in the monkey ventral striatum and putamen. PloS One. 2015;10(6):e0130443.
71. Haber SN, Knutson B. The reward circuit: Linking primate anatomy and human imaging. Neuropsychopharmacology. 2009;35(1):4.
72. Shi Z, Ma Y, Wu B, et al. Neural correlates of reflection on actual versus ideal self-discrepancy. NeuroImage. 2016;124(Part A):573; Spreckelmeyer KN, Krach S, Kohls G, et al. Anticipation of monetary and social reward differently activates mesolimbic brain structures in men and women. Social Cognitive and Affective Neuroscience. 2009;4(2):158–65; Stelly CE, Pomrenze MB, Cook JB, Morikawa H. Repeated social defeat stress enhances glutamatergic synaptic plasticity in the VTA and cocaine place conditioning. Elife. 2016;5:e15448(1–18).
73. Asaad, Eskandar. Encoding of both positive and negative reward prediction errors.
74. Gale JT, Shields DC, Ishizawa Y, Eskandar EN. Reward and reinforcement activity in the nucleus accumbens during learning. Frontiers in Behavioral Neuroscience. 2014;8:114(1–10); Fontanini A, Grossman S, Figueroa J, Katz D. Distinct subtypes of basolateral amygdala taste neurons reflect palatability and reward. Journal of Neuroscience. 2009;29(8):2486–95.

3. THE UNIVERSE OF HUMAN REWARDS

1. Flood MM. Some experimental games. Management Science. 1958;5(1):5–26; Kahneman D, Knetsch JL, Thaler RH. Fairness and the assumptions of economics. (Proceedings from a conference held October 13–15, 1985, at the University of Chicago). Journal of Business. 1986;59(4):S285; Camerer CF, Fehr E. Measuring social norms and preferences using experimental games: a guide for social scientists. In: Henrich J, Boyd R, Bowles S, et al., editors. Foundations of human sociality: economic experiments and ethnographic evidence from fifteen small-scale societies. New York: Oxford University Press; 2004, pp. 55–95.
2. Locey ML, Safin V, Rachlin H. Social discounting and the prisoner's dilemma game. Journal of the Experimental Analysis of Behavior. 2013;99(1):85–97; Camerer CF.

Strategizing in the brain. Science. 2003;300(5626):1673; Camerer C, Loewenstein G, Prelec D. Neuroeconomics: How neuroscience can inform economics. Journal of Economic Literature. 2005;43(1):9–64; Hetzer M, Sornette D. The co-evolution of fairness preferences and costly punishment. PLoS One. 2013;8(3):e54308.

3. Madani K. Modeling international climate change negotiations more responsibly: can highly simplified game theory models provide reliable policy insights? Ecological Economics. 2013;90:68–76; Decanio S, Fremstad A. Game theory and climate diplomacy. Ecological Economics. 2013;85:177–87; Soroos M. Global change, environmental security, and the prisoner's dilemma. Journal of Peace Research. 1994;31:317.

4. Henrich J, Smith N. Comparative experiments evidence from Machiguenga, Mapuche, Huinca, and American populations. In: Henrich, Boyd, Bowles, et al., editors. Foundations of human sociality: economic experiments and ethnographic evidence from fifteen small-scale societies. New York: Oxford University Press; 2004, pp. 125–67; Marlowe F. Dictators and ultimatums in an egalitarian society of hunter-gatherers, the Hadza of Tanzania. In: Henrich, Boyd, Bowles, et al., editors. Foundations of human sociality, pp. 168–93.

5. Proctor D, Williamson RA, de Waal FBM, Brosnan SF. Chimpanzees play the ultimatum game. (Author abstract). Proceedings of the National Academy of Sciences of the United States. 2013;110(6):2070.

6. Pillutla M, Murnighan J. Unfairness, anger, and spite: emotional rejections of ultimatum offers. Organizational Behavior & Human Decision Processes. 1996; 68(3):208–24.

7. Sanfey AG, Rilling JK, Aronson JA, et al. The neural basis of economic decision-making in the ultimatum game. Science 2003;300(5626):1755.

8. Gospic K, Mohlin E, Fransson P, et al. Limbic justice—amygdala involvement in immediate rejection in the ultimatum game. PLoS Biology. 2011;9(5):e1001054.

9. Hetzer M, Sornette D. The co-evolution of fairness preferences and costly punishment. PLoS One. 2013;8(3):e54308.

10. D'Ardenne K, McClure SM, Nystrom LE, Cohen JD. BOLD responses reflecting dopaminergic signals in the human ventral tegmental area. Science. 2008;319(5867):1264.

11. Hart AS, Rutledge RB, Glimcher PW, Phillips PEM. Phasic dopamine release in the rat nucleus accumbens symmetrically encodes a reward prediction error term. Journal of Neuroscience. 2014;34(3):698.

12. Masayuki M, Okihide H. Two types of dopamine neuron distinctly convey positive and negative motivational signals. Nature. 2009;459(7248):837; Asaad WF, Eskandar EN. Encoding of both positive and negative reward prediction errors by neurons of the primate lateral prefrontal cortex and caudate nucleus. Journal of Neuroscience. 2011;31(49):17772.

13. Haber SN, Knutson B. The reward circuit: linking primate anatomy and human imaging. Neuropsychopharmacology. 2009;35(1):4; Kishida KT, Saez I, Lohrenz T, et al. Subsecond dopamine fluctuations in human striatum encode superposed error signals about actual and counterfactual reward. Proceedings of the National Academy of Sciences of the United States of America. 2016;113(1):200.
14. Dohmen T, Falk A, Fliessbach K, et al. Relative versus absolute income, joy of winning, and gender: brain imaging evidence. Journal of Public Economics. 2011;95(3):279–85; Fliessbach K, Weber B, Trautner P, et al. Social comparison affects reward-related brain activity in the human ventral striatum. Science. 2007;318(5854): 1305.
15. Dohmen, Falk, Fliessbach, et al. Relative versus absolute income, joy of winning, and gender.
16. Massi B, Luhmann C. Fairness influences early signatures of reward-related neural processing. Cognitive, Affective, & Behavioral Neuroscience. 2015;15(4):768–75; Chang SWC, Fagan NA, Toda K, et al. Neural mechanisms of social decision-making in the primate amygdala. Proceedings of the National Academy of Sciences of the United States of America. 2015;112(52):16012; Chang SWC, Gariépy J-F, Platt ML. Neuronal reference frames for social decisions in primate frontal cortex. Nature Neuroscience. 2012;16(2):243.
17. Seo H, Lee D. Temporal filtering of reward signals in the dorsal anterior cingulate cortex during a mixed-strategy game. Journal of Neuroscience. 2007;27(31):8366–77.
18. Haber, Knutson. The reward circuit.
19. Sterling P. Why we consume: Neural design and sustainability. Great transition initiative. 2016. http://www.greattransition.org/publication/why-we-consume.
20. Dawkins R. The selfish gene. Thirtieth anniversary edition. Oxford: Oxford University Press; 2006; Manner M, Gowdy J. The evolution of social and moral behavior: evolutionary insights for public policy. Ecological Economics. 2010;69(4):753–761.
21. Yamagishi T, Mifune N, Li Y, et al. Is behavioral pro-sociality game-specific? Pro-social preference and expectations of pro-sociality. Organizational Behavior & Human Decision Processes. 2013;120(2):260.
22. Tricomi E, Rangel A, Camerer CF, O'Doherty JP. Neural evidence for inequality-averse social preferences. Nature. 2010;463(7284):1089.
23. Saez I, Zhu L, Set E, et al. Dopamine modulates egalitarian behavior in humans. Current Biology. 2015;25(7):912–19.
24. Zizzo DJ, Tan JHW. Game harmony: a behavioral approach to predicting cooperation in games. American Behavioral Scientist. 2011;55(8):987–1013.
25. Karsh N, Eitam B. I control therefore I do: judgments of agency influence action selection. Cognition. 2015;138:122–31.
26. Eitam B, Kennedy P, Higgins E. Motivation from control. Experimental Brain Research. 2013;229(3):475–84.

27. Eitam, Kennedy, Higgins. Motivation from control.

28. DePasque Swanson S, Tricomi E. Goals and task difficulty expectations modulate striatal responses to feedback. Cognitive, Affective, & Behavioral Neuroscience. 2014;14(2):610–20.

29. Tarbell IM. The history of the Standard Oil Company. New York: McClure, Phillips & Company; 1904.

30. Alcaro A, Panksepp J, Huber R. D-amphetamine stimulates unconditioned exploration / approach behaviors in crayfish: towards a conserved evolutionary function of ancestral drug reward. Pharmacology, Biochemistry and Behavior. 2011;99(1):75–80.

31. Wetherford MJ, Cohen LB. Developmental changes in infant visual preferences for novelty and familiarity. Child Development. 1973;44(3):416–24; Rose SA, Gottfried AW, Melloy-Carminar P, Bridger WH. Familiarity and novelty preferences in infant recognition memory: implications for information processing. Developmental Psychology. 1982;18(5):704–13; Colombo J, Bundy RS. Infant response to auditory familiarity and novelty. Infant Behavior and Development. 1983;6(2):305–11.

32. Horvitz JC. Mesolimbocortical and nigrostriatal dopamine responses to salient non-reward events. Neuroscience. 2000;96(4):651–56; Bunzeck N, Doeller CF, Fuentemilla L, Dolan RJ, Duzel E. Reward motivation accelerates the onset of neural novelty signals in humans to 85 milliseconds. Current Biology. 2009;19(15):1294–1300; Lawson AL, Liu X, Joseph J, et al. Sensation seeking predicts brain responses in the old–new task: converging multimodal neuroimaging evidence. International Journal of Psychophysiology. 2012;84(3):260–69.

33. Hart A, Clark J, Phillips P. Dynamic shaping of dopamine signals during probabilistic Pavlovian conditioning. Neurobiology of Learning and Memory. 2015;117:84–92; Budygin E, Park J, Bass CE, et al. Aversive stimulus differentially triggers subsecond dopamine release in reward regions. Neuroscience. 2012;201:331–37; Kishida K, Sandberg SG, Lohrenz T, et al. Sub-second dopamine detection in human striatum. PLoS One. 2011;6(8):200–205; Kishida KT, Saez I, Lohrenz T, et al. Subsecond dopamine fluctuations in human striatum encode superposed error signals about actual and counterfactual reward. Proceedings of the National Academy of Sciences of the United States of America. 2016;113(1):200; Krebs RM, Schott BH, Düzel E. Personality traits are differentially associated with patterns of reward and novelty processing in the human substantia nigra / ventral tegmental area. Biological Psychiatry. 2009;65(2):103–10.

34. Bromberg-Martin ES, Matsumoto M, Hikosaka O. Dopamine in motivational control: rewarding, aversive, and alerting. Neuron. 2010;68(5):815–34.

35. Sterling, Why we consume.

36. Clark CA, Dagher A. The role of dopamine in risk taking: a specific look at Parkinson's disease and gambling. Frontiers in Behavioral Neuroscience. 2014;8(article 196):1–12.

37. Dreher JC, Kohn P, Kolachana B, et al. Variation in dopamine genes influences responsivity of the human reward system. Proceedings of the National Academy of

Sciences of the United States of America. 2009;106(2):617–22; Birkas E, Horváth J, Lakatos K, et al. Association between dopamine D4 receptor (DRD4) gene polymorphisms and novelty-elicited auditory event-related potentials in preschool children. Brain Research. 2006;1103(1):150–58.

38. Lawson, Liu, Joseph, et al. Sensation seeking predicts brain responses in the old–new task; Krebs, Schott, Düzel. Personality traits are differentially associated with patterns of reward and novelty processing; Schwartz C, Wright C, Shin L, et al. Inhibited and uninhibited infants "grown up": adult amygdalar response to novelty. Science. 2003;300(5627):1952–53.

39. Vargas-López V, Torres-Berrio A, González-Martínez L, et al. Acute restraint stress and corticosterone transiently disrupts novelty preference in an object recognition task. Behavioural Brain Research. 2015; 291:60–66.

40. Wetzel N, Widmann A, Schröger E. Processing of novel identifiability and duration in children and adults. Biological Psychology. 2011;86(1):39–49.

41. Mather E, Plunkett K. The role of novelty in early word learning. Cognitive Science. 2012;36(7):1157–77.

42. Quilty LC, Oakman JM, Farvolden P. Behavioural inhibition, behavioural activation, and the preference for familiarity. Personality and Individual Differences. 2007;42(2): 291–303; Takeshi S, Masato I. Examining familiarity through the temperament and character inventory: a structural equation modeling analysis. Behaviormetrika. 2011;38(2):139–51.

43. Schwabe I, Jonker W, Berg S. Genes, culture and conservatism—a psychometric-genetic approach. Behavioural Genetics. 2016;46(4):516–28.

44. Eaves L, Heath A, Martin N, et al. Comparing the biological and cultural inheritance of personality and social attitudes in the Virginia 30 000 study of twins and their relatives. Twin Research and Human Genetics. 1999;2(2):62–80; Hatemi P, Medland S, Morley K, Heath A, Martin N. The genetics of voting: an Australian twin study. Behavioural Genetics. 2007;37(3):435–48; Schwabe, Jonker, Berg. Genes, culture and conservatism; Hibbing JR, Smith KB, Alford JR. Differences in negativity bias underlie variations in political ideology. Behavioral and Brain Sciences. 2014;37(3):297–307; Weeden J, Kurzban R. Do people naturally cluster into liberals and conservatives? Evolutionary Psychological Science. 2016;2(1):47–57.

45. Hibbing, Smith, Alford. Differences in negativity bias underlie variations in political ideology; Dennis TA, Amodio DM, O'Toole LJ. Associations between parental ideology and neural sensitivity to cognitive conflict in children. Social Neuroscience. 2014; 10(02):206–17; Williams LM, Gatt JM, Grieve SM, et al. COMT Val108 / 158Met polymorphism effects on emotional brain function and negativity bias. NeuroImage. 2010;53(3):918–25.

46. Hornsey M, Harris E, Bain P, Fielding K. Meta-analyses of the determinants and outcomes of belief in climate change. Nature Climate Change. 2016;6(6):622–26.

47. Val-Laillet D, Meurice P, Clouard C. Familiarity to a feed additive modulates its effects on brain responses in reward and memory regions in the pig model. PLoS One. 2016;11(9):e0162660.

48. Scheele D, Wille A, Kendrick KM, et al. Oxytocin enhances brain reward system responses in men viewing the face of their female partner. Proceedings of the National Academy of Sciences of the United States of America. 2013;110(50):20308.

49. Blaustein JD. Neuroendocrine regulation of feminine sexual behavior: lessons from rodent models and thoughts about humans. Annual Review of Psychology. 2008;59:93; Campbell BC, Dreber A, Apicella CL, et al. Testosterone exposure, dopaminergic reward, and sensation-seeking in young men. Physiology & Behavior. 2010;99(4):451–56; Gerra G, Avanzini P, Zaimovic A, et al. Neurotransmitters, neuroendocrine correlates of sensation-seeking temperament in normal humans. Neuropsychobiology. 1999;39(4):207–13.

50. Acevedo BP, Aron A, Fisher HE, Brown LL. Neural correlates of marital satisfaction and well-being: reward, empathy, and affect. Clinical Neuropsychiatry: Journal of Treatments Evaluation. 2012;9(1):20.

51. Henrich, Boyd, Bowles, et al., editors. Foundations of human sociality; Marlowe F. Dictators and ultimatums in an egalitarian society of hunter-gatherers, the Hadza of Tanzania. In: Henrich, Boyd, Bowles, et al., editors. Foundations of human sociality, pp. 168–193.

52. Lawrence EA. Feline fortunes: Contrasting views of cats in popular culture. Journal of Popular Culture. 2003;36(3):623–35.

53. Lawrence, Feline fortunes.

54. Dawkins R. *The god delusion*. London: Bantam Press; 2006; Losin E, Woo C-W, Krishnan A, et al. Brain and psychological mediators of imitation: sociocultural versus physical traits. Cult Brain. 2015;3(2):93–111.

55. Lee K, Cameron CA, Doucette J, Talwar V. Phantoms and fabrications: young children's detection of implausible lies. Child Development. 2002;73(6):1688–1702.

56. Hitchens C. God is not great: how religion poisons everything. First edition. New York: Hachette, 2007; Dawkins, The God Delusion.

57. Ariely D, Loewenstein G, Prelec D. Tom Sawyer and the construction of value. Journal of Economic Behavior and Organization. 2006;60(1):1–10.

4. BIOPHILIA AND THE BRAIN

1. Fromm E. The heart of man: its genius for good and evil. 1st ed. New York: Harper & Row; 1964.

2. Wilson EO. Biophilia. Cambridge, MA: Harvard University Press; 1984.

3. Wells NM, Lekies KS. Nature and the life course: pathways from childhood nature experiences to adult environmentalism. Children Youth and Environments. 2006;16(1):1–24.

4. Diamond J. New Guineans and their natural world. In: Kellert SR, Wilson EO, editors. The biophilia hypothesis. Mercer Island, WA: Island Books; 1993, pp. 251–71; Bratman GN, Hamilton JP, Daily GC. The impacts of nature experience on human cognitive function and mental health. Year in Ecology and Conservation Biology. 2012;1249(1):118–36.

5. Ulrich RS. Biophilia, biophobia, and natural landscapes. In: Kellert SR, Wilson EO, editors. The biophilia hypothesis. Washington, DC: Island Press, Shearwater Books; 1993, pp. 73–137; Kaplan S, Kaplan R, Wendt J. Rated preference and complexity for natural and urban visual material. Perception & Psychophysics. 1972;12(4):354–56; Mahidin AMM, Maulan S. Understanding children's preferences of natural environment as a start for environmental sustainability. Procedia—Social and Behavioral Sciences. 2012;38:324–33.

6. Ulrich RS. Human responses to vegetation and landscapes. Landscape and Urban Planning. 1986;13:29–44; Balling JD, Falk JH. Development of visual preference for natural environments. Environment and Behavior. 1982;14(1):5–28.

7. Ulrich. Human responses to vegetation and landscapes.

8. Ulrich. Biophilia, biophobia, and natural landscapes, pp. 73–137; Fletcher JH, Emlen JA. Comparison of the responses to snakes of lab- and wild-reared rhesus monkeys. Animal Behaviour. 1964;12(2):348–52.

9. Öhman A, Mineka S. Fears, phobias, and preparedness: toward an evolved module of fear and fear learning. Psychological Review. 2001;108(3):483–522.

10. Ulrich. Biophilia, biophobia, and natural landscapes, pp. 73–137.

11. Fjørtoft I. The natural environment as a playground for children: the impact of outdoor play activities in pre-primary school children. Early Childhood Education Journal. 2001;29(2):111–17; Atchley RA, Strayer DL, Atchley P. Creativity in the wild: improving creative reasoning through immersion in natural settings. (Research article). PLoS One. 2012;7(12):e51474; Shin DJ. ARTICLES: The effects of changes in outdoor play environment on children's cognitive and social play behaviors. International Journal of Early Childhood Education. 1998;3:77; Burdette HL, Whitaker RC. Resurrecting free play in young children: looking beyond fitness and fatness to attention, affiliation, and affect. Archives of Pediatrics & Adolescent Medicine. 2005;159(1):46–50; Gray P. Play theory of hunter-gatherer egalitarianism. In: Narvaez D, Valentino K, Fuentes A, et al., editors. Ancestral landscapes in human evolution: culture, childrearing, and social wellbeing. New York: Oxford University Press; 2014, pp. 192–215; Kahn PH. Developmental psychology and the biophilia hypothesis: children's affiliation with nature. Developmental Review. 1997;17(1):1–61.

12. Fjørtoft I, Sageie J. The natural environment as a playground for children: landscape description and analyses of a natural playscape. Landscape and Urban Planning. 2000;48(1):83–97; Herrington S, Studtmann K. Landscape interventions: new directions for the design of children's outdoor play environments. Landscape and Urban Planning. 1998;42(2):191–205.

13. Herrington, Studtmann. Landscape interventions.

14. Gray P. Play theory of hunter-gatherer egalitarianism. In: Narvaez D, Valentino K, Fuentes A, et al., editors. Ancestral landscapes in human evolution: culture, childrearing, and social wellbeing. New York: Oxford University Press; 2014. pp. 192–215; Gray P. Free to learn: why unleashing the instinct to play will make our children happier, more self-reliant, and better students for life. New York: Basic Books; 2013; Lew-Levy S, Reckin R, Lavi N, et al. How do hunter-gatherer children learn subsistence skills? Human Nature 2017;28(4):367–94; Singer DG, Singer JL, D'Agnostino H, DeLong R. Children's pastimes and play in sixteen nations: is free-play declining? American Journal of Play. 2009;1(3):283–312; Hrdy SB. Mothers and others: the evolutionary origins of mutual understanding. Cambridge, MA: The Belknap Press of Harvard University Press; 2009.

15. Wells N, Evans G. Nearby nature: a buffer of life stress among rural children. Environment and Behavior. 2003;35(3):311–30; Newell PB. A cross-cultural examination of favorite places. Environment and Behavior. 1997;29(4):495.

16. Norðdahl K, Einarsdóttir J. Children's views and preferences regarding their outdoor environment. Journal of Adventure Education and Outdoor Learning. 2015;15(2):152–67; Kirby MA. Nature as refuge in children's environments. Children's Environments Quarterly. 1989;6(1):7–12.

17. Berto R. Exposure to restorative environments helps restore attentional capacity. Journal of Environmental Psychology. 2005;25(3):249–59.

18. Louv R. Last child in the woods: saving our children from nature-deficit disorder. Updated and expanded. ed. Chapel Hill, NC: Algonquin Books of Chapel Hill; 2008.

19. van Den Berg AE, van Den Berg CG. A comparison of children with ADHD in a natural and built setting. Child: Care, Health and Development. 2011;37(3):430; Kuo FE, Tayler AF. A potential natural treatment for attention-deficit / hyperactivity disorder: evidence from a national study. American Journal of Public Health. 2004;94:1580–86; Berry M, Sweeney M, Morath J, et al. The nature of impulsivity: visual exposure to natural environments decreases impulsive decision-making in a delay discounting task. PLoS One. 2014;9(5):e97915.

20. Matsuoka RH. Student performance and high school landscapes: examining the links. Landscape and Urban Planning. 2010;97(4):273–82.

21. Kuo F, Sullivan W. Aggression and violence in the inner city: effects of environment via mental fatigue. Environment and Behavior. 2001;33(4):543–71.

22. Ulrich RS, Simons RF, Losito BD, et al. Stress recovery during exposure to natural and urban environments. Journal of Environmental Psychology. 1991;11:201–30; Purcell T, Peron E, Berto R. Why do preferences differ between scene types? Environment and Behavior. 2001;33(1):93–106; van Den Berg M, Maas J, Muller R, et al. Autonomic nervous system responses to viewing green and built settings: differentiating between sympathetic and parasympathetic activity. International Journal of Environmental Research and Public Health. 2015;12(12):15860–74.

23. Ulrich, Simons, Losito, et al. Stress recovery during exposure to natural and urban environments.

24. McAllister TW, Flashman LA, McDonald BC, Saykin AJ. Mechanisms of working memory dysfunction after mild and moderate TBI: evidence from functional MRI and neurogenetics. Journal of Neurotrauma. 2006;23(10):1450.

25. Hartig T, Mang M, Evans GW. Restorative effects of natural environment experiences. Environment and Behavior. 1991;23(1):3–26; Lee KE, Williams KJH, Sargent LD, et al. 40-second green roof views sustain attention: the role of micro-breaks in attention restoration. Journal of Environmental Psychology. 2015;42:182–89.

26. Katcher A, Wilkins G. Dialogue with animals: its nature and culture. In: Kellert SR, Wilson EO, editors. The biophilia hypothesis. Washington, DC: Shearwater Books; 1993. pp. 173–97; Ward A, Arola N, Bohnert A, Lieb R. Social-emotional adjustment and pet ownership among adolescents with autism spectrum disorder. (Report). Journal of Communication Disorders. 2017;65:35.

27. Wood L, Martin K, Hayley C, et al. The pet factor—companion animals as a conduit for getting to know people, friendship formation and social support. PLoS One. 2015;10(4):e0122085; Herzog H. The impact of pets on human health and psychological well-being: fact, fiction, or hypothesis? Current Directions in Psychological Science. 2011;20(4):236–39.

28. Herzog H. The impact of pets on human health and psychological well-being: fact, fiction, or hypothesis? Current Directions in Psychological Science. 2011;20(4):236–39; Thorpe R, Simonsick E, Brach J, et al. Dog ownership, walking behavior, and maintained mobility in late life. Journal of the American Geriatrics Society. 2006;54(9): 1419–24; Himsworth CG, Rock M. Pet ownership, other domestic relationships, and satisfaction with life among seniors: results from a Canadian National Survey. Anthrozoös. 2013;26(2):295–305; Batty GD, Zaninotto P, Watt RG, Bell S. Associations of pet ownership with biomarkers of ageing: population based cohort study. BMJ. 2017;359.

29. Grinde B, Patil GG. Biophilia: does visual contact with nature impact on health and well-being? International Journal of Environmental Research and Public Health. 2009;6(9):2332–43.

30. Ulrich RS. View through a window may influence recovery from surgery. Science. 1984;224(4647):420–21.

31. Park S-H, Mattson RH. Effects of flowering and foliage plants in hospital rooms on patients recovering from abdominal surgery. HortTechnology. 2008;18(4):563–68; Park S, Mattson RH. Therapeutic influences of plants in hospital rooms on surgical recovery. Hortscience. 2009;44(1):102–5.

32. Raanaas RK, Patil GG, Hartig T. Health benefits of a view of nature through the window: a quasi-experimental study of patients in a residential rehabilitation center. Clinical Rehabilitation. 2012;26(1):21–32.

33. Wilker EH, Wu C-D, McNeely E, et al. Green space and mortality following ischemic stroke. Environmental Research. 2014;133:42–48.

34. Donovan GH, Butry DT, Michael YL, et al. The relationship between trees and human health: evidence from the spread of the emerald ash borer. American Journal of Preventive Medicine. 2012;44(2):139–145.

35. Gullone E. The biophilia hypothesis and life in the 21st century: increasing mental health or increasing pathology? Journal of Happiness Studies. 2000;1:293–321.

36. Bratman GN, Hamilton JP, Daily GC. The impacts of nature experience on human cognitive function and mental health. Year in Ecology and Conservation Biology. 2012;1249(1):118–36; Dzhambov AM, Dimitrova DD. Elderly visitors of an urban park, health anxiety and individual awareness of nature experiences. Urban Forestry & Urban Greening. 2014;13(4):806–13.

37. Hunter MR, Gillespie BW, Chen SY-P. Urban nature experiences reduce stress in the context of daily life based on salivary biomarkers. Frontiers in Psychology. 2019;10.

38. Grahn P, Palsdottir A, Ottosson J, Jonsdottir I. Longer nature-based rehabilitation may contribute to a faster return to work in patients with reactions to severe stress and / or depression. International Journal of Environmental Research and Public Health. 2017;14(11); Clatworthy J, Hinds J, Camic P. Gardening as a mental health intervention: a review. Mental Health Review. 2013;18(4):214–25; Renzetti C, Follingstad D. From blue to green: the development and implementation of a therapeutic horticulture program for residents of a battered women's shelter. Violence and Victims. 2015;30(4):676–90.

39. Jiler J. Doing time in the garden: life lessons through prison horticulture. Cannizzo J, editor. Oakland, CA: New Village Press; 2006; Ulrich C, Nadkarni N. Sustainability research and practices in enforced residential institutions: collaborations of ecologists and prisoners. Environment, Development and Sustainability. 2009;11(4):815–32; Brown G, Bos E, Brady G, et al. An evaluation of the Master Gardener Programme at HMP Rye Hill: A horticultural intervention with substance misusing offenders. Prison Service Journal. 2016(225):45; O'Callaghan AM, Robinson ML, Reed C, Roof L. Horticultural training improves job prospects and sense of well being for prison inmates. Acta Horticulturae. 2010(8812):773–78; Polomski RF, Johnson KM, Anderson JC. Prison inmates become master gardeners in South Carolina. HortTechnology. 1997(4):360–62; Robinson ML, O'Callaghan AM. Expanding horticultural training into the prison population. Journal of Extension. 2008;46(4).

40. South EC, Hohl BC, Kondo MC, et al. Effect of greening vacant land on mental health of community-dwelling adults. JAMA Network Open. 2018;1(3):e180298.

41. Dadvand P, Villanueva CM, Font-Ribera L, et al. Risks and benefits of green spaces for children: a cross-sectional study of associations with sedentary behavior, obesity, asthma, and allergy. Environmental Health Perspectives 2014;122(12):1329–35.

42. Bangsbo J, Krustrup P, Duda J, et al. The Copenhagen Consensus Conference 2016: children, youth, and physical activity in schools and during leisure time. British

Journal of Sports Medicine. 2016(50):1177–78; McCurdy LE, Winterbottom KE, Mehta SS, Roberts JR. Using nature and outdoor activity to improve children's health. Current Problems in Pediatric and Adolescent Health Care. 2010;40(5):102–17.

43. Wells N. At home with nature: effects of "greenness" on children's cognitive functioning. Environment and Behavior. 2000;32(6):775–95.

44. Woodgate RL, Skarlato O. "It is about being outside": Canadian youth's perspectives of good health and the environment. Health Place. 2015;31:100–10; Gopinath B, Hardy LL, Baur LA, et al. Physical activity and sedentary behaviors and health-related quality of life in adolescents. Pediatrics. 2012;130(1):e167.

45. Wiens V, Kyngäs H, Pölkki T. The meaning of seasonal changes, nature, and animals for adolescent girls' wellbeing in northern Finland: a qualitative descriptive study. International Journal of Qualitative Studies on Health and Well-Being. 2016;11(1): 30160–14.

46. Berry, Sweeney, Morath, et al. The nature of impulsivity.

47. Engemann K, Pedersen CB, Arge L, et al. Residential green space in childhood is associated with lower risk of psychiatric disorders from adolescence into adulthood. Proceedings of the National Academy of Sciences of the United States of America. 2019;116(11):5188.

48. Gidlow CJ, Jones MV, Hurst G, et al. Where to put your best foot forward: psychophysiological responses to walking in natural and urban environments. Journal of Environmental Psychology. 2016;45:22–29.

49. de Vries S, Verheij RA, Groenewegen PP, Spreeuwenberg P. Natural environments—healthy environments? An exploratory analysis of the relationship between greenspace and health. Environment and Planning A. 2003;35(10):1717–31.

50. Gong Y, Gallacher J, Palmer S, Fone D. Neighbourhood green space, physical function and participation in physical activities among elderly men: the Caerphilly Prospective study. International Journal of Behavioral Nutrition and Physical Activity. 2014;11:40; Takano T, Nakamura K, Watanabe M. Urban residential environments and senior citizens' longevity in megacity areas: the importance of walkable green spaces. Journal of Epidemiology and Community Health. 2002;56(12):913; Wang D, Lau KK-L, Yu R, et al. Neighbouring green space and mortality in community-dwelling elderly Hong Kong Chinese: a cohort study. BMJ Open. 2017;7(7):e015794-e.

51. Huynh Q, Craig W, Janssen I, Pickett W. Exposure to public natural space as a protective factor for emotional well-being among young people in Canada. (Research article) BMC Public Health. 2013;13:407; Saw L, Lim F, Carrasco L. The relationship between natural park usage and happiness does not hold in a tropical city-state. PLoS One. 2015;10(7):e0133781.

52. Calogiuri G. Natural environments and childhood experiences promoting physical activity, examining the mediational effects of feelings about nature and social networks. International Journal of Environmental Research and Public Health. 2016;13(4).

53. Adams S, Savahl S. Children's perceptions of the natural environment: a South African perspective. Children's Geographies. 2013(2):196–211; Chong S, Lobb E, Khan R, et al. Neighbourhood safety and area deprivation modify the associations between parkland and psychological distress in Sydney, Australia. (Research article) BMC Public Health. 2013;13:422.
54. Joye Y, De Block A. 'Nature and I are Two': a critical examination of the biophilia hypothesis. Environmental Values. 2011;20(2):189–215.
55. Diamond. New Guineans and their natural world; Tierney K, Connolly M. A review of the evidence for a biological basis for snake fears in humans. Psychological Record. 2013;63(4):919–28.
56. Adams, Savahl. Children's perceptions of the natural environment; Balling JD, Falk JH. Development of visual preference for natural environments. Environment and Behavior. 1982;14(1):5–28; Saw, Lim, Carrasco. The relationship between natural park usage and happiness does not hold in a tropical city-state.
57. Fischer CS. The biophilia hypothesis. 1994. p. 1161.
58. Adams, Savahl. Children's perceptions of the natural environment.
59. Kahn PH. Developmental psychology and the biophilia hypothesis: children's affiliation with nature. Developmental Review. 1997;17(1):1–61.
60. Louv R. Last child in the woods: saving our children from nature-deficit disorder. Updated and expanded ed. Chapel Hill, NC: Algonquin Books of Chapel Hill; 2008.
61. Albrecht G, Sartore G-M, Connor L, et al. The distress caused by environmental change. Australasian Psychiatry. 2007;15(1 suppl):S95–S8; Walton T, Shaw WS. Living with the Anthropocene blues. Geoforum. 2015;60:1–3.
62. Cunsolo Willox A, Stephenson E, Allen J, et al. Examining relationships between climate change and mental health in the Circumpolar North. Regional Environmental Change. 2015;15(1):169–82; Ellis NR, Albrecht GA. Climate change threats to family farmers' sense of place and mental wellbeing: a case study from the Western Australian wheatbelt. Social Science & Medicine. 2017;175:161–68; Rice SM, McIver LJ. Climate change and mental health: rationale for research and intervention planning. Asian Journal of Psychiatry. 2016;20:1–2.
63. Gifford E, Gifford R. The largely unacknowledged impact of climate change on mental health. Bulletin of the Atomic Scientists. 2016;72(5):292–97.
64. Zeyer A, Roth W-M. Post-ecological discourse in the making. Public Understanding of Science. 2013;22(1):33–48; Majeed H, Lee J. The impact of climate change on youth depression and mental health. The Lancet Planetary Health. 2017;1(3):e94–e95; Van Den Hazel P. Perspective on children's public mental health and climate change. European Journal of Public Health. 2017;27.
65. Walton T, Shaw WS. Living with the Anthropocene blues. Geoforum. 2015;60:1–3.
66. Gifford, Gifford. The largely unacknowledged impact of climate change on mental health.

67. Verplanken B, Roy D. "My worries are rational, climate change is not"; habitual ecological worrying is an adaptive response. PLoS One. 2013;8(9):e74708.

5. AN ACCELERATION OF CONSUMPTION

1. Hámori J. History of human brain evolution. Frontiers of Neuroscience. 2010;4; Leigh SR. Brain growth, life history, and cognition in primate and human evolution. American Journal of Primatology. 2004;62(3):139–64; Antón S, Potts R, Aiello L. Evolution of early *Homo:* An integrated biological perspective. Science. 2014;345(6192): 1236828.
2. Maguire EA, Spiers HJ, Good CD, et al. Navigation expertise and the human hippocampus: a structural brain analysis. Hippocampus. 2003;13:250–59; Maguire EA, Woollett K, Spiers HJ. London taxi drivers and bus drivers: a structural MRI and neuropsychological analysis. Hippocampus. 2006;16:1091–101.
3. Meilinger T, Frankenstein J, Bülthoff HH. Learning to navigate: experience versus maps. Cognition. 2013;129(1):24–30; Meilinger T, Frankenstein J, Simon N, et al. Not all memories are the same: situational context influences spatial recall within one's city of residency. Psychon Bull Rev. 2016;23(1):246–52; Frankenstein J. Is GPS all in our heads? New York Times. 2012 February 5, 2012.
4. Gindrat A-D, Chytiris M, Balerna M, et al. Use-dependent cortical processing from fingertips in touchscreen phone users. Current Biology. 2015;25(1):109–16.
5. Akers K, Martinez-Canabal A, Restivo L, et al. Hippocampal neurogenesis regulates forgetting during adulthood and infancy. Science. 2014;344(6184):598–602.
6. Parker ES, Cahill L, McGaugh JL. A case of unusual autobiographical remembering. Neurocase. 2006;12(1):35–49; Leport AKR, Stark SM, McGaugh JL, Stark CEL. A cognitive assessment of highly superior autobiographical memory. Memory. 2017;25(2):276–88.
7. Sterling P, Laughlin S. Principles of neural design. Cambridge, MA: MIT Press; 2015.
8. Bae B-I, Tietjen I, Atabay KD, et al. Evolutionarily dynamic alternative splicing of GPR56 regulates regional cerebral cortical patterning. Science. 2014;343(6172):764; Rash B, Rakic P. Genetic resolutions of brain convolutions. Science. 2014;343(6172):744–45; Konopka G, Geschwind DH. Human brain evolution: harnessing the genomics (r)evolution to link genes, cognition, and behavior. Neuron. 2010;68(2):231–44.
9. Iriki A, Taoka M. Triadic (ecological, neural, cognitive) niche construction: a scenario of human brain evolution extrapolating tool use and language from the control of reaching actions. Philosophical Transactions of the Royal Society B. 2012;367(1585): 10–23.
10. Iriki, Taoka. Triadic (ecological, neural, cognitive) niche construction.
11. Toffler A. Future shock. New York: Random House; 1970.

12. Mika P. Future shock—discussing the changing temporal architecture of daily life. Journal of Futures Studies. 2010;14(4):1–21.

13. Burger O, Baudisch A, Vaupel JW. Human mortality improvement in evolutionary context. Proceedings of the National Academy of Sciences of the United States of America. 2012;109(44):18210–14.

14. Allen JG, MacNaughton P, Satish U, et al. Associations of cognitive function scores with carbon dioxide, ventilation, and volatile organic compound exposures in office workers: A controlled exposure study of green and conventional office environments. Environmental Health Perspectives. 2015;124(6):805–812.

15. Hofferth S, Sandberg J. Changes in American children's time, 1981–1997. Advances in Life Course Research. 2001;6:193–229.

16. Anderson DR, Huston AC, Schmitt KL, et al. Early childhood television viewing and adolescent behavior: The recontact study. Monographs of the Society for Research in Child Development. 2001;66(1):1–147.

17. Vandewater EA, Rideout VJ, Wartella EA, et al. Digital childhood: electronic media and technology use among infants, toddlers, and preschoolers. Pediatrics. 2007;119(5):e1006.

18. Singer DG, Singer JL, D'Agnostino H, DeLong R. Children's pastimes and play in sixteen nations: is free-play declining? American Journal of Play. 2009;1(3):283–312.

19. Jago R, Stamatakis E, Gama A, et al. Parent and child screen-viewing time and home media environment. American Journal of Preventive Medicine. 2012;43(2): 150–58; Tandon PS, Zhou C, Lozano P, Christakis DA. Preschoolers' total daily screen time at home and by type of child care. Journal of Pediatrics. 2011;158(2): 297–300.

20. Jago, Stamatakis, Gama, et al. Parent and child screen-viewing time; Lauricella AR, Wartella E, Rideout VJ. Young children's screen time: the complex role of parent and child factors. Journal of Applied Developmental Psychology. 2015;36:11–7; Atkin AJ, Corder K, van Sluijs EMF. Bedroom media, sedentary time and screen-time in children: a longitudinal analysis. International Journal of Behavioral Nutrition and Physical Activity. 2013;10:137; Ramirez ER, Norman GJ, Rosenberg DE, et al. Adolescent screen time and rules to limit screen time in the home. Journal of Adolescent Health. 2011;48(4):379–85.

21. Yan Z. Child and adolescent use of mobile phones: an unparalleled complex developmental phenomenon. Child Development. 2018;89(1):5–16.

22. Hiniker A, Sobel K, Suh H, et al. Texting while parenting: how adults use mobile phones while caring for children at the playground. Proceedings of the 33rd Annual ACM (Association for Computing Machinery) Conference on human factors on computing systems. ACM Digital Media. 2015:727–36; Kushlev K, Dunn EW. Smartphones distract parents from cultivating feelings of connection when spending time with their children. Journal of Social and Personal Relationships. 2018;36(6):1619–39;

Radesky JS, Kistin CJ, Zuckerman B, et al. Patterns of mobile device use by caregivers and children during meals in fast food restaurants. Pediatrics. 2014;133(4):e843.

23. Brod C. Managing technostress: optimizing the use of computer technology. Personnel Journal. 1982;61(10):753–57; Lee Y, Chang CT, Lin Y, Cheng Z. The dark side of smartphone usage: Psychological traits, compulsive behavior and technostress. Computers in Human Behavior. 2014;31:373–83.

24. Chassiakos Y, Radesky J, Christakis D, et al. Children and adolescents and digital media. Pediatrics. 2016;138(5):e20162593.

25. Chassiakos, Radesky, Christakis, et al. Children and adolescents and digital media; Divan HA, Kheifets L, Obel C, Olsen J. Cell phone use and behavioural problems in young children. Journal of Epidemiology and Community Health. 2012;66(6):524; Hinkley T, Verbestel V, Ahrens W, et al. Early childhood electronic media use as a predictor of poorer well-being: a prospective cohort study. JAMA Pediatrics. 2014;168(5):485–92; Jackson LA, Von Eye A, Fitzgerald HE, et al. Internet use, videogame playing and cell phone use as predictors of children's body mass index (BMI), body weight, academic performance, and social and overall self-esteem. Computers in Human Behavior. 2011;27:599–604; Yan. Child and adolescent use of mobile phones; Brown A, Smolenaers E. Parents' interpretations of screen time recommendations for children younger than 2 years. Journal of Family Issues. 2018;39(2):406–29.

26. Tene O, Polonetsky J. A theory of creepy: technology, privacy and shifting social norms. Yale Journal of Law and Technology. 2014;16(1):59–102.

27. Sterling P, Laughlin S. Principles of neural design. Cambridge, MA: MIT Press; 2015.

28. Hawkins RP, Yong-Ho K, Pingree S. The ups and downs of attention to television. Communication Research. 1991;18(1):53–76.

29. Anderson DR, Levin SR. Young children's attention to "Sesame Street." Child Development. 1976;47(3):806–11.

30. Sproull N. Visual attention, modeling behaviors, and other verbal and nonverbal meta-communication of prekindergarten children viewing Sesame Street. American Educational Research Journal. 1973;10(2):101–14.

31. Anderson DR, Levin SR. Young children's attention to "Sesame Street"; Hawkins RP, Yong-Ho K, Pingree S. The ups and downs of attention to television.

32. Christakis D, Zimmerman F, Digiuseppe D, McCarty C. Early television exposure and subsequent attentional problems in children. Pediatrics. 2004;113(4):708–13.

33. Chassiakos, Radesky, Christakis, et al. Children and adolescents and digital media.

34. Cristia A, Seidl A. Parental reports on touch screen use in early childhood. PLoS One. 2015;10(6):p.e0128338-e0128338; DOI:10.1371 / journal.pone.0128338.

35. Korkeamäki R-L, Dreher MJ, Pekkarinen A. Finnish preschool and first-grade children's use of media at home. Human Technology: An Interdisciplinary Journal on

Humans in ICT Environments. 2012;8(2):109–32; Chassiakos, Radesky, Christakis, et al. Children and adolescents and digital media.

36. Chassiakos, Radesky, Christakis, et al. Children and adolescents and digital media.
37. Bavelier D, Green CS, Dye M. Children, wired: for better and for worse. Neuron. 2010;67(5):692–701.
38. Lauricella, Wartella, Rideout. Young children's screen time; Korkeamäki, Dreher, Pekkarinen. Finnish preschool and first-grade children's use of media.
39. Séguin D, Klimek V. Just five more minutes please: electronic media use, sleep and behaviour in young children. Early Child Development and Care. 2015;186(6):981–1000; Roser K, Schoeni A, Roosli M. Mobile phone use, behavioural problems and concentration capacity in adolescents: a prospective study. International Journal of Hygiene and Environmental Health. 2016;219(8):759–69; Taehtinen RE, Sigfusdottir ID, Helgason AR, Kristjansson AL. Electronic screen use and selected somatic symptoms in 10–12 year old children. Preventive Medicine. 2014;67:128–33; Divan, Kheifets Obel, Olsen. Cell phone use and behavioural problems; Lemola S, Perkinson-Gloor N, Brand S, et al. Adolescents' electronic media use at night, sleep disturbance, and depressive symptoms in the smartphone age. Journal of Youth and Adolescence. 2015;44(2):405–18; Pecor K, Kang L, Henderson M, et al. Sleep health, messaging, headaches, and academic performance in high school students. Brain & Development. 2016;38(6):548–53.
40. Tomopoulos S, Dreyer B, Berkule S, et al. Infant media exposure and toddler development. Archives of Pediatrics & Adolescent Medicine. 2010;164(12):1105–11.
41. Hinkley, Verbestel, Ahrens, et al. Early childhood electronic media use.
42. Jackson, Von Eye, Fitzgerald, et al. Internet use, videogame playing and cell phone use as predictors.
43. Yuan K, Cheng P, Dong T, Bi Y, et al. Cortical thickness abnormalities in late adolescence with online gaming addiction. PLoS ONE. 2013;8(1):e53055; Zhu Y, Zhang H, Tian M. Molecular and functional imaging of internet addiction. Biomed Research International. 2015;2015:378675–9.
44. Hinkley T, Cliff DP, Okely AD. Reducing electronic media use in 2–3 year-old children: feasibility and efficacy of the Family@play pilot randomised controlled trial. (Report). BMC Public Health. 2015;15(1):779.
45. Anderson, Huston, Schmitt, et al. Early childhood television viewing and adolescent behavior.
46. Rodgers RF, Damiano SR, Wertheim EH, Paxton SJ. Media exposure in very young girls: prospective and cross-sectional relationships with BMIz, self-esteem and body size stereotypes. Developmental Psychology. 2017;53(12):2356–63; Anderson, Huston, Schmitt, et al. Early childhood television viewing and adolescent behavior; Stamou AG, Maroniti K, Griva E. Young children talk about their popular cartoon and TV

heroes' speech styles: media reception and language attitudes. Language Awareness. 2015;24(3):1–17.

47. Bavelier, Green, Dye. Children, wired; Christakis DA, Garrison MM, Herrenkohl T, et al. Modifying media content for preschool children: a randomized controlled trial. (Report). Pediatrics. 2013;131(3):431; Wilson BJ. Media and children's aggression, fear, and altruism. The Future of Children. 2008;18(1):87–118; Anderson, Huston, Schmitt, et al. Early childhood television viewing and adolescent behavior.

48. Mathiak KA, Klasen M, Weber R, et al. Reward system and temporal pole contributions to affective evaluation during a first person shooter video game. (Report). BMC Neuroscience. 2011;12:66.

49. Chassiakos, Radesky, Christakis, et al. Children and adolescents and digital media.

50. Christakis, Garrison, Herrenkohl, et al. Modifying media content for preschool children; Wilson. Media and children's aggression, fear, and altruism.

51. Rogoff B, Morelli GA, Chavajay P. Children's integration in communities and segregation from people of differing ages. Perspectives on Psychological Science. 2010;5(4):431–40.

52. Anderson KJ, Cavallaro D. Parents or pop culture? Children's heroes and role models. Childhood Education. 2002;78(3):161–68; Chen-Yu JH, Seock Y-K. Adolescents' clothing purchase motivations, information sources, and store selection criteria: a comparison of male / female and impulse / nonimpulse shoppers. Family and Consumer Sciences Research Journal. 2002;31(1):50–77.

53. Bogin B. Childhood, adolescence, and longevity: a multilevel model of the evolution of reserve capacity in human life history. American Journal of Human Biology. 2009;21(4):567–77; Bogin B, Varea C, Hermanussen M, Scheffler C. Human life course biology: a centennial perspective of scholarship on the human pattern of physical growth and its place in human biocultural evolution. American Journal of Physical Anthropology. 2018;165(4):834–54.

54. Lebel C, Beaulieu C. Longitudinal development of human brain wiring continues from childhood into adulthood. Journal of Neuroscience. 2011;31(30):10937–47; Spear LP. Adolescent neurodevelopment. Journal of Adolescent Health. 2013;52(2):S7–S13; Galvan A, Hare T, Voss H, et al. Risk-taking and the adolescent brain: who is at risk? Developmental Science. 2007;10(2):F8–F14; Steinberg L. A social neuroscience perspective on adolescent risk-taking. Developmental Review. 2008;28(1):78–106; Yeatman JD, Wandell BA, Mezer AA. Lifespan maturation and degeneration of human brain white matter. Nature Communcations 2014;5:4932.

55. Duhaime A-C. Imitation, immaturity, and injury. Journal of Neurosurgery Pediatrics. 2009;4(5):407.

56. Patton GC, Olsson CA, Skirbekk V, et al. Adolescence and the next generation. Nature. 2018;554(7693):458.

57. Chassiakos, Radesky, Christakis, et al. Children and adolescents and digital media.
58. Ehrenreich S, Underwood M, Ackerman R. Adolescents' text message communication and growth in antisocial behavior across the first year of high school. Journal of Abnormal Child Psychology. 2014;42(2):251–64; Anderson L, McCabe DB. A coconstructed world: adolescent self-socialization on the internet. (Report). Journal of Public Policy & Marketing. 2012;31(2):240; Romer D, Moreno M. Digital media and risks for adolescent substance abuse and problematic gambling. Pediatrics. 2017;140 (Suppl 2):S102.
59. Stamou, Maroniti, Griva. Young children talk about their popular cartoon and TV heroes' speech styles.
60. Davies P, Surridge J, Hole L, Munro-Davies L. Superhero-related injuries in paediatrics: a case series. Archives of Disease in Childhood. 2007;92(3):242.
61. Anderson, Cavallaro. Parents or pop culture?; Holub S, Tisak M, Mullins D. Gender differences in children's hero attributions: Personal hero choices and evaluations of typical male and female heroes. Sex Roles. 2008;58(7):567–78; Gash H, Rodríguez P. Young people's heroes in France and Spain. Spanish Journal of Psychology. 2009;12(1):246–57.
62. Anderson, McCabe. A coconstructed world; Sato N, Kato Y. Youth marketing in Japan. Young Consumers. 2005;6(4):56; Montgomery K. Youth and digital media: a policy research agenda. Journal of Adolescent Health. 2000;27(2):61–68; Shim S, Serido J, Barber B. A consumer way of thinking: linking consumer socialization and consumption motivation perspectives to adolescent development. Journal of Research on Adolescence. 2011;21(1):290–99; Bax T. Internet addiction in China: The battle for the hearts and minds of youth. Deviant Behavior. 2014;35(9):687–702.
63. Hefner D, Knop K, Schmitt S, Vorderer P. Rules? Role model? Relationship? The impact of parents on their children's problematic mobile phone involvement. Media Psychology. 2018:22(1):82–108; Tur-Porcar A. Parenting styles and internet use. Psychology & Marketing. 2017;34(11):1016–22.
64. Kurniawan S, Haryanto J. Kids as future market: the role of autobiographical memory in building brand loyalty. Researchers World. 2011;2(4):77–90; Confos N, Davis T. Young consumer–brand relationship building potential using digital marketing. European Journal of Marketing. 2016;50(11):1993–2017; Sharma A, Sonwaney V. Theoretical modeling of influence of children on family purchase decision making. Procedia—Social and Behavioral Sciences. 2014;133:38–46; Kaur P, Singh R. Children in family purchase decision making in India and the West. Academy of Marketing Science Review. 2006;2006(8):1–30; Chen-Yu, Seock. Adolescents' clothing purchase motivations, information sources, and store selection criteria; de Vries L, Gensler S, Leeflang PSH. Popularity of brand posts on brand fan pages: an investigation of the effects of social media marketing. Journal of Interactive Marketing. 2012;26(2):83–91; McClure A, Tanski S, Li Z, et al. Internet alcohol marketing and underage alcohol use. Pediatrics. 2016;137(2):p.e20152149–e20152149; Soneji S, Pierce JP, Choi K, et al. Engagement with online tobacco marketing and associations with tobacco product

use among US youth. Journal of Adolescent Health. 2017;61(1):61–69; Hill WE, Beatty SE, Walsh G. A segmentation of adolescent online users and shoppers. Journal of Services Marketing. 2013;27(5):347–60; Sato, Kato. Youth marketing in Japan.

65. Lawlor M-A, Dunne Á, Rowley J. Young consumers' brand communications literacy in a social networking site context. European Journal of Marketing. 2016;50(11):2018–40; Holmberg CE. Chaplin J, Hillman T, Berg C. Adolescents' presentation of food in social media: an explorative study. Appetite. 2016;99(C):121–29; An S, Jin HS, Park EH. Children's advertising literacy for advergames: Perception of the game as advertising. Journal of Advertising. 2014;43(1):63–72; Okazaki S. The tactical use of mobile marketing: how adolescents' social networking can best shape brand extensions. Journal of Advertising Research. 2009;49(1):12; Montgomery KC, Chester J. Interactive food and beverage marketing: Targeting adolescents in the digital age. Journal of Adolescent Health. 2009;45(3):S18–S29; Niu HJ, Chiang YS, Tsai HT. An exploratory study of the Otaku adolescent consumer. Psychology & Marketing. 2012;29(10):712–25.

66. Opel DJ, Diekema DS, Lee NR, Marcuse EK. Social marketing as a strategy to increase immunization rates. Archives of Pediatrics & Adolescent Medicine. 2009;163(5):432–37; Yan. Child and adolescent use of mobile phones; de Vreese CH. Digital renaissance: young consumer and citizen? Annals of the American Academy of Political and Social Science. 2007;611(1):207–16; Wilson. Media and children's aggression, fear, and altruism.

67. Nelson JK, Zeckhauser R. The patron's payoff: conspicuous commissions in Italian Renaissance art. Princeton, NJ: Princeton University Press; 2008.

68. Kaur, Singh. Children in family purchase decision making in India and the West; Sener A. Influences of adolescents on family purchasing behavior: perceptions of adolescents and parents. Social Behavior and Personality. 2011;39(6):747–54; Marshall R, Reday PA. Internet-enabled youth and power in family decisions. Young Consumers. 2007;8(3):177–83.

69. Sharma, Sonwaney. Theoretical modeling of influence of children on family purchase decision making; Rose G, Boush D, Shoham A. Family communication and children's purchasing influence: a cross-national examination. Journal of Business Research. 2002;55(11):867–73; Lawlor M-A, Prothero A. Pester power—a battle of wills between children and their parents. Journal of Marketing Management. 2011;27(5–6):561.

70. Chen-Yu, Seock. Adolescents' clothing purchase motivations, information sources, and store selection criteria.

71. Chassiakos, Radesky, Christakis, et al. Children and adolescents and digital media.

72. Winpenny EM, Marteau TM, Nolte E. Exposure of children and adolescents to alcohol marketing on social media websites. Alcohol and Alcoholism. 2014;49(2):154–59; Barrientos-Gutiérrez T, Barrientos-Gutiérrez I, Reynales-Shigematsu LM, et al. Se busca mercado adolescente: internet y videojuegos, las nuevas estrategias de la industria tabacalera. Salud publica de Mexico. 2012;54(3):303.

73. Zwarun L, Linz D, Metzger M, Kunkel D. Effects of showing risk in beer commercials to young drinkers. Journal of Broadcasting & Electronic Media. 2006;50(1):52–77.

74. Bernhardt A, Wilking C, Gilbert-Diamond D, et al. Children's recall of fast food television advertising—testing the adequacy of food marketing regulation. PLoS One. 2015;10(3):e0119300.

75. Tene O, Polonetsky J. A theory of creepy: technology, privacy and shifting social norms. Yale Journal of Law and Technology. 2014;16(1):59–102.

76. Chen-Yu, Seock. Adolescents' clothing purchase motivations, information sources, and store selection criteria.

77. Minahan S, Huddleston P. Shopping with my mother: reminiscences of adult daughters. International Journal of Consumer Studies. 2013;37(4):373–78.

78. Garcia JR, Saad G. Evolutionary neuromarketing: Darwinizing the neuroimaging paradigm for consumer behavior. Journal of Consumer Behaviour. 2008;7(4–5):397–414; Saad G, Vongas JG. The effect of conspicuous consumption on men's testosterone levels. Organizational Behavior and Human Decision Processes. 2009;110(2):80–92; Saad G. The consuming instinct. What Darwinian consumption reveals about human nature. Politics and the Life Sciences: Journal of the Association for Politics and the Life Sciences. 2013;32(1):58; Griskevicius V, Kenrick DT. Fundamental motives: How evolutionary needs influence consumer behavior. Journal of Consumer Psychology. 2013;23(3):372–86.

79. Hayhoe CR, Leach L, Turner PR. Discriminating the number of credit cards held by college students using credit and money attitudes. Journal of Economic Psychology. 1999;20(6):643–56.

80. Claes L, Müller A, Norré J, et al. The relationship among compulsive buying, compulsive internet use and temperament in a sample of female patients with eating disorders. European Eating Disorders Review. 2012;20(2):126–31; Lawrence LM, Ciorciari J, Kyrios M. Relationships that compulsive buying has with addiction, obsessive-compulsiveness, hoarding, and depression. Comprehensive Psychiatry. 2014;55(5):1137–45; Loredana V, Gioacchino L, Santo DIN. Buying addiction: reliability and construct validity of an assessment questionnaire. Postmodern Openings. 2015;VI(1):149–60; Mueller A, Mitchell JE, Peterson LA, et al. Depression, materialism, and excessive internet use in relation to compulsive buying. Comprehensive Psychiatry. 2011;52(4):420.

81. Sterling P. Why we consume: neural design and sustainability: great transition initiative; 2016. [Available from: http://www.greattransition.org/publication/why-we-consume.]

82. Hajdu G, Hajdu T. The impact of culture on well-being: evidence from a natural experiment. Journal of Happiness Studies. 2015;17(3):1089–1110.

83. Vaillant GE. Triumphs of experience: the men of the Harvard Grant Study. Cambridge, MA: Belknap Press of Harvard University Press; 2012.

84. Veenhoven R, Diener E, Michalos A. Editorial: what this journal is about. Journal of Happiness Studies. 2000;1(1):5–8.

85. Kim EJ, Kyeong S, Cho SW, et al. Happier people show greater neural connectivity during negative self-referential processing. PLOS One. 2016;11(2): e0149554–e0149554; Heller AS, van Reekum CM, Schaefer SM, et al. Sustained striatal activity predicts eudaimonic well-being and cortisol output. Psychological Science. 2013;24(11):2191; Luo S, Yu D, Han S. Genetic and neural correlates of romantic relationship satisfaction. Social Cognitive and Affective Neuroscience. 2016;11(2):337–48; Shi Z, Ma Y, Wu B, et al. Neural correlates of reflection on actual versus ideal self-discrepancy. NeuroImage. 2016;124(Part A):573.

86. Kong F, Hu S, Wang X, et al. Neural correlates of the happy life: the amplitude of spontaneous low frequency fluctuations predicts subjective well-being. NeuroImage. 2015;107:136–45; Kong F, Wang X, Hu S, Liu J. Neural correlates of psychological resilience and their relation to life satisfaction in a sample of healthy young adults. NeuroImage. 2015;123:165.

87. Mrazek MD, Mooneyham BW, Mrazek KL, Schooler JW. Pushing the limits: cognitive, affective, and neural plasticity revealed by an intensive multifaceted intervention. Frontiers in Human Neuroscience. 2016;10:117–117.

6. WHICH BEHAVIORS MATTER MOST

1. Schrag DP. Geobiology of the Anthropocene. In: Knoll AH, Canfield DE, Konhauser KO, editors. Fundamentals of geobiology: Oxford: Blackwell Publishing; 2012.

2. McKibben B. A special moment in history, part 2: Earth 2. The Atlantic Monthly. 1998;281(5):55–78.

3. Hertwich E, Peters G. Carbon footprint of nations: a global, trade-linked analysis. Environmental Science & Technology. 2009;43(16):6414; World Bank. CO_2 emissions (metric tons per capita) 2013. [Available at: https://data.worldbank.org/indicator/EN.ATM.CO_2E.PC.]

4. Oreskes N, Conway EM. Merchants of doubt: how a handful of scientists obscured the truth on issues from tobacco smoke to global warming. New York: Bloomsbury Press; 2010.

5. Environmental Protection Agency. Global greenhouse gas emissions data 2017 [updated April 13, 2017]. [Available at: https://www.epa.gov/ghgemissions/global-greenhouse-gas-emissions-data]; MacKay DJC. Sustainable energy without the hot air. Cambridge: UIT Cambridge; 2009.

6. Solomon S, Plattner GK, Knutti R, Friedlingstein P. Irreversible climate change due to carbon dioxide emissions. Proceedings of the National Academy of Sciences of the United States of America. 2009;106(6):1704–9.

7. Bin S, Dowlatabadi H. Consumer lifestyle approach to US energy use and the related CO_2 emissions. Energy Policy. 2005;33(2):197–208.

8. Bin, Dowlatabadi. Consumer lifestyle approach to US energy use.
9. Environmental Protection Agency. Global greenhouse gas emissions data 2017.
10. Chung JW, Meltzer DO. Estimate of the carbon footprint of the US health care sector. JAMA. 2009;302(18):1970–72; Eckelman MJ, Sherman J. Environmental impacts of the U.S. health care system and effects on public health. (Report). PLoS One. 2016;11(6): e0157014–e0157014.
11. MacKay. Sustainable energy without the hot air; Stern PC. Psychology and the science of human–environment interactions. American Psychologist. 2000;55(5):523–30.
12. Bin, Dowlatabadi. Consumer lifestyle approach to US energy use; Jones CM, Kammen DM. Quantifying carbon footprint reduction opportunities for U.S. households and communities. Environmental Science & Technology. 2011;45(9):4088–95; Hertwich, Peters. Carbon footprint of nations.
13. Stern. Psychology and the science of human–environment interactions; MacKay. Sustainable energy without the hot air.
14. Dietz T, Gardner GT, Gilligan J, et al. Household actions can provide a behavioral wedge to rapidly reduce US carbon emissions. Proceedings of the National Academy of Sciences of the United States of America. 2009;106(44):18452; Schnoor JL. Coalitions of the willing. Environmental Science & Technology. 2012;46(17):9201.
15. Gössling S, Cohen S. Why sustainable transport policies will fail: EU climate policy in the light of transport taboos. Journal of Transport Geography. 2014;39:197–207.
16. Pacala S, Socolow R. Stabilization wedges: solving the climate problem for the next 50 years with current technologies. Science. 2004;305(5686):968–72.
17. Intergovernmental Panel on Climate Change. Intergovernmental Panel on Climate Change Website 2017 [Available at: http://www.ipcc.ch/]; Intergovernmental Panel on Climate Change. IPCC 2018: Summary for policymakers. In: Masson-Delmotte V, Zhai P, Pörtner HO, et al., editors. Global warming of 15°C. An IPCC special report on the impacts of global warming of 15°C above pre-industrial levels and related global greenhouse gas emission pathways, in the context of strengthening the global response to the threat of climate change, sustainable development, and efforts to eradicate poverty. Geneva: World Meteorologic Organization; 2018. pp. 1–32.
18. McKibben. A special moment in history, part 2.
19. Schrag. Geobiology of the Anthropocene; Siegenthaler U, Stocker TF, Monnin E, et al. Stable carbon cycle-climate relationship during the Late Pleistocene. Science. 2005; 310(5752):1313.
20. Intergovernmental Panel on Climate Change. IPCC, 2013: summary for policymakers. In: Stocker TF, Qin D, Plattner GK, et al., editors. Climate Change 2013: The physical science basis contribution of Working Group I to the Fifth Assessment Report of the Intergovernmental Panel on Climate Change. Cambridge: Cambridge University Press; 2013. pp. 3–29.

21. Borie M, Mahony M, Obermeister N, Hulme M. Knowing like a global expert organization: comparative insights from the IPCC and IPBES. Global Environmental Change. 2021;68:102261.

22. Girod B, van Vuuren DP, Hertwich EG. Global climate targets and future consumption level: an evaluation of the required ghg intensity. Environmental Research Letters. 2013;8(1):014016.

23. Harris N, Payne O, Mann SA. How much rainforest is in that chocolate bar? World Resources Institute; 2015. [Available at: http://www.wri.org/blog/2015/08/how-much-rainforest-chocolate-bar.]

24. Recanati F, Marveggio D, Dotelli G. From beans to bar: a life cycle assessment towards sustainable chocolate supply chain. Science of the Total Environment. 2018;613–14: 1013–23.

25. Mackey B. Counting trees, carbon and climate change. Significance. 2014;11(1):19–23; Williamson P. Scrutinize CO_2 removal methods: the viability and environmental risks of removing carbon dioxide from the air must be assessed if we are to achieve the Paris goals. Nature. 2016;530(7589):153.

26. Gillingham K, Sweeney J. Barriers to implementing low-carbon technologies. Climate Change Economics. 2012;3(04):1250019; Williamson. Scrutinize CO_2 removal methods; Goeppert A, Czaun M, May RB, et al. Carbon dioxide capture from the air using a polyamine based regenerable solid adsorbent. Journal of the American Chemical Society. 2011;133(50):20164.

27. Attari SZ, Dekay ML, Davidson CI, de Bruin WB. Public perceptions of energy consumption and savings. Proceedings of the National Academy of Sciences of the United States of America. 2010;107(37):16054; Whitmarsh L. Behavioural responses to climate change: asymmetry of intentions and impacts. Journal of Environmental Psychology. 2009;29(1):13–23.

28. Wynes S, Nicholas KA. The climate mitigation gap: education and government recommendations miss the most effective individual actions. Environmental Research Letters. 2017;12(7):074024.

29. Gardner GT, Stern PC. The short list: the most effective actions U.S. households can take to curb climate change. Environment: Science and Policy for Sustainable Development. 2008;50(5):12–25.

30. Stern. Psychology and the science of human–environment interactions; Girod B, van Vuuren DP, Hertwich EG. Climate policy through changing consumption choices: options and obstacles for reducing greenhouse gas emissions. Global Environmental Change. 2014;25(March 2014):5–15.

31. World Bank. CO_2 emissions.

32. Environmental Protection Agency. Global greenhouse gas emissions data 2017; MacKay. Sustainable energy without the hot air.

33. Ehrlich PR, Holdren JP. Impact of population growth. Science. 1971;171(3977):1212–17; Schrag. Geobiology of the Anthropocene; Girod, van Vuuren, Hertwich. Global climate targets and future consumption level.

34. Girod B, van Vuuren DP, Deetman S. Global travel within the 2°C climate target. Energy Policy. 2012;45:152–66; Intergovernmental Panel on Climate Change. IPCC, 2013: summary for policymakers. In: Stocker, Qin, Plattner, et al., editors. Climate change 2013; Garren S, Pinjari A, Brinkmann R. Carbon dioxide emission trends in cars and light trucks: a comparative analysis of emissions and methodologies for Florida's counties (2000 and 2008). Energy Policy. 2011;39(9):5287; Hao H, Geng Y, Sarkis J. Carbon footprint of global passenger cars: scenarios through 2050. Energy. 2016;101:121–31.

35. Fuglestvedt J, Berntsen T, Myhre G, et al. Climate forcing from the transport sectors. Proceedings of the National Academy of Sciences of the United States of America. 2008;105(2):454; Gössling, Cohen. Why sustainable transport policies will fail.

36. Hao, Geng, Sarkis. Carbon footprint of global passenger cars.

37. Wynes, Nicholas. The climate mitigation gap.

38. Gardner, Stern. The short list.

39. Reichmuth DS, Lutz AE, Manley DK, Keller JO. Comparison of the technical potential for hydrogen, battery electric, and conventional light-duty vehicles to reduce greenhouse gas emissions and petroleum consumption in the United States. International Journal of Hydrogen Energy. 2012(38):1200–1208.

40. Fladung A-K, Gron G, Grammer K, et al. A neural signature of anorexia nervosa in the ventral striatal reward system. American Journal of Psychiatry. 2010;167(2):206.

41. Ensslen A, Schücking M, Jochem P, et al. Empirical carbon dioxide emissions of electric vehicles in a French-German commuter fleet test. Journal of Cleaner Production. 2017;142:263–78.

42. Bellekom S, Benders R, Pelgröm S, Moll H. Electric cars and wind energy: two problems, one solution? A study to combine wind energy and electric cars in 2020 in the Netherlands. Energy. 2012;45(1):859–66.

43. Wanitschke A, Hoffmann S. Are battery electric vehicles the future? An uncertainty comparison with hydrogen and combustion engines. Environmental Innovation and Societal Transitions. 2020;35:509–23.

44. Singh B, Guest G, Bright R, Strømman A. Life cycle assessment of electric and fuel cell vehicle transport based on forest biomass. Journal of Industrial Ecology. 2014;18(2):176–86; Hao H, Qiao Q, Liu Z, Zhao F. Impact of recycling on energy consumption and greenhouse gas emissions from electric vehicle production: the China 2025 case. Resources, Conservation & Recycling. 2017;122:114–25.

45. Holdway A, Williams A, Inderwildi O, King D. Indirect emissions from electric vehicles: emissions from electricity generation. Energy & Environmental Science.

2010;3(12):1825–32; Singh, Guest, Bright, Strømman. Life cycle assessment of electric and fuel cell vehicle transport.

46. Hao, Geng, Sarkis. Carbon footprint of global passenger cars.
47. Oreskes N, Conway EM. Defeating the merchants of doubt. Nature. 2010;465:686–87.
48. Girod, van Vuuren, Deetman. Global travel within the 2°C climate target; Edwards HA, Dixon-Hardy D, Wadud Z. Aircraft cost index and the future of carbon emissions from air travel. Applied Energy. 2016;164:553–62; Lee DS, Pitari G, Grewe V, et al. Transport impacts on atmosphere and climate: aviation. Atmospheric Environment. 2010;44(37):4678–734.
49. Lee, Pitari, Grewe, et al. Transport impacts on atmosphere and climate.
50. Edwards, Dixon-Hardy, Wadud. Aircraft cost index and the future of carbon emissions from air travel.
51. Loo BPY, Li L. Carbon dioxide emissions from passenger transport in China since 1949: implications for developing sustainable transport. Energy Policy. 2012;50:464–76.
52. Hayward JA, O' Connell DA, Raison RJ, et al. The economics of producing sustainable aviation fuel: a regional case study in Queensland, Australia. GCB Bioenergy. 2015;7(3): 497–511; Lee, Pitari, Grewe, et al. Transport impacts on atmosphere and climate.
53. Grote M, Williams I, Preston J. Direct carbon dioxide emissions from civil aircraft. Atmospheric Environment. 2014;95:214–24.
54. Rosenthal E. Your biggest carbon sin may be air travel. New York Times. January 27, 2013.
55. Wynes, Nicholas. The climate mitigation gap; Gardner, Stern. The short list.
56. Stohl A. The travel-related carbon dioxide emissions of atmospheric researchers. Atmospheric Chemistry and Physics. 2008;8(21):6499–504.
57. Girod, van Vuuren, Deetman. Global travel within the 2° C climate target.
58. Jakob M. Marginal costs and co-benefits of energy efficiency investments: the case of the Swiss residential sector. Energy Policy. 2006;34(2):172–87.
59. Gardner, Stern. The short list.
60. Girod, van Vuuren, Hertwich. Climate policy through changing consumption choices.
61. Gardner, Stern. The short list; Dietz, Gardner, Gilligan, et al. Household actions can provide a behavioral wedge.
62. Hertwich EG, Roux C. Greenhouse gas emissions from the consumption of electric and electronic equipment by Norwegian households. Environmental Science & Technology. 2011;45(19):8190.
63. Scott MJ, Dirks JA, Cort KA. The value of energy efficiency programs for U.S. residential and commercial buildings in a warmer world. Mitigation and Adaptation Strategies for Global Change. 2007;13(4):307–339.
64. Tilman D, Clark M. Global diets link environmental sustainability and human health. Nature. 2014;515(7528):518–522.

65. Pollan M. The omnivore's dilemma: a natural history of four meals. New York: Penguin Press; 2006; Pollan M. Food rules: an eater's manual. New York: Penguin Press; 2009; Beavan C. No impact man: the adventures of a guilty liberal who attempts to save the planet, and the discoveries he makes about himself and our way of life in the process. 1st ed. New York: Farrar, Strauss, and Giroux; 2009.

66. Tilman, Clark. Global diets link environmental sustainability and human health.

67. Eshel G, Shepon A, Makov T, Milo R. Land, irrigation water, greenhouse gas, and reactive nitrogen burdens of meat, eggs, and dairy production in the United States. Proceedings of the National Academy of Sciences of the United States of America. 2014;111(33):11996–2001; Pan A, Sun Q, Bernstein AM, et al. Red meat consumption and mortality: results from 2 prospective cohort studies. Archives of Internal Medicine. 2012;172(7):555–63.

68. Tilman, Clark. Global diets link environmental sustainability and human health.

69. Nemecek T, Jungbluth N, Canals L, Schenck R. Environmental impacts of food consumption and nutrition: where are we and what is next? International Journal of Life Cycle Assessment. 2016;21(5):607–20; Heller MC, Keoleian G, Willett W. Toward a life cycle-based, diet-level framework for food environmental impact and nutritional quality assessment: a critical review. Environmental Science & Technology. 2013; 47(22):12632–47.

70. Nemecek, Jungbluth, Canals, Schenck. Environmental impacts of food consumption and nutrition; Brodt S, Kramer K, Kendall A, Feenstra G. Comparing environmental impacts of regional and national-scale food supply chains: a case study of processed tomatoes. Food Policy. 2013;42:106–14.

71. Garnett T. Food sustainability: problems, perspectives and solutions. Proceedings of the Nutrition Society. 2013;72(1):29–39.

72. Lukas M, Rohn H, Lettenmeier M, et al. The nutritional footprint—integrated methodology using environmental and health indicators to indicate potential for absolute reduction of natural resource use in the field of food and nutrition. Journal of Cleaner Production. 2016;132:161–70.

73. Tilman, Clark. Global diets link environmental sustainability and human health.

74. Wynes, Nicholas. The climate mitigation gap.

7. BEHAVIORS THAT ARE EASY AND HARD TO CHANGE

1. Balleine BW, Dezfouli A. Hierarchical action control: adaptive collaboration between actions and habits. Frontiers in Psychology. 2019;10:2735.

2. Yin HH, Knowlton BJ. The role of the basal ganglia in habit formation. Nature Reviews Neuroscience. 2006;7(6):464–76; Smith KS, Graybiel AM. Habit formation coincides with shifts in reinforcement representations in the sensorimotor striatum. Journal of Neurophysiology. 2016;115(3):1487.

3. Brembs B. Spontaneous decisions and operant conditioning in fruit flies. Behavioural Processes. 2011;87(1):157–64; Ena S, D'Exaerde AD, Schiffmann SN. Unraveling the differential functions and regulation of striatal neuron sub-populations in motor control, reward, and motivational processes. Frontiers in Behavioral Neuroscience 2011(5)47:1–10; Faure A, Haberland U, Conde F, El Massioui N. Lesion to the nigrostriatal dopamine system disrupts stimulus-response habit formation. Journal of Neuroscience. 2005;25(11):2771–80; Humphries MD, Prescott TJ. The ventral basal ganglia, a selection mechanism at the crossroads of space, strategy, and reward. Progress in Neurobiology. 2010;90:385–417.

4. Swim J, Clayton S, Doherty T, et al. Psychology and global climate change: addressing a multi-faceted phenomenon and set of challenges. A report by the American Psychological Association's Task Force on the Interface between Psychology and Global Climate Change: American Psychological Association; 2011. [Available from: http://www.apa.org/science/about/publications/climate-change.aspx.]

5. Morgan D. Schedules of reinforcement at 50: a retrospective appreciation. The Psychological Record. 2010;60(1):151–72.

6. Ferster CB, Skinner BF. Schedules of reinforcement. New York: Appleton-Century-Crofts; 1957; Sterling P. Why we consume: neural design and sustainability: Great Transition Initiative; 2016. [Available from: http://www.greattransition.org/publication/why-we-consume.]

7. Torrubia R, Ávila C, Moltó J, Caseras X. The Sensitivity to Punishment and Sensitivity to Reward Questionnaire (SPSRQ) as a measure of Gray's anxiety and impulsivity dimensions. Personality and Individual Differences. 2001;31(6):837–62.

8. Morgan. Schedules of reinforcement at 50; Hardin MG, Perez-Edgar K, Guyer AE, et al. Reward and punishment sensitivity in shy and non-shy adults: relations between social and motivated behavior. Personality and Individual Differences. 2006;40(4):699–711.

9. Peterson RF, Peterson LR. The use of positive reinforcement in the control of self-destructive behavior in a retarded boy. Journal of Experimental Child Psychology. 1968;6(3):351–60; Boyd LA, Keilbaugh WS, Axelrod S. The direct and indirect effects of positive reinforcement on on-task behavior. Behavior Therapy. 1981;12(1):80–92; Andreou TE, McIntosh K, Ross SW, Kahn JD. Critical incidents in sustaining school-wide positive behavioral interventions and supports. Journal of Special Education. 2015;49(3):157–67; Dickinson DJ. Changing behavior with behavioral techniques. Journal of School Psychology. 1968;6(4):278–83; Payne SW, Dozier CL. Positive reinforcement as treatment for problem behavior maintained by negative reinforcement. Journal of Applied Behavior Analysis. 2013;46(3):699–703; Manassis K, Young A. Adapting positive reinforcement systems to suit child temperament. Journal of the American Academy of Child and Adolescent Psychiatry. 2001;40(5):603–605; Slocum SK, Vollmer TR. A comparison of positive and negative reinforcement for compliance to treat problem behavior maintained by escape. Journal of Applied Behavior Analysis. 2015;48(3):563–74; Call NA, Lomas Mevers JE. The relative influence of motivating

operations for positive and negative reinforcement on problem behavior during demands: MOs for positive versus negative reinforcement. Behavioral Interventions. 2014;29(1):4–20.

10. Gabor AM, Fritz JN, Roath CT, et al. Caregiver preference for reinforcement-based interventions for problem behavior maintained by positive reinforcement. Journal of Applied Behavior Analysis. 2016;49(2):215–27.

11. Adams K, Heath D, Sood G, et al. Positive reinforcement to create and sustain a culture change in hand hygiene practices in a tertiary care academic facility. American Journal of Infection Control. 2013;41(6):S95.

12. Prochaska JJ, Prochaska JO. A review of multiple health behavior change interventions for primary prevention. American Journal of Lifestyle Medicine. 2011;5(3):208–21; Johnston W, Buscemi J, Coons M. Multiple health behavior change: a synopsis and comment on "A review of multiple health behavior change interventions for primary prevention." Translational Behavioral Medicine. 2013;3(1):6–7; Harrison A, Newell M-L, Imrie J, Hoddinott G. HIV prevention for South African youth: which interventions work? A systematic review of current evidence. BMC Public Health. 2010;10:102.

13. Ventikos N, Lykos G, Padouva I. How to achieve an effective behavioral-based safety plan: the analysis of an attitude questionnaire for the maritime industry. WMU Journal of Maritime Affairs. 2014;13(2):207–30.

14. Sims B. Using positive-reinforcement programs to effect culture change. Employment Relations Today. 2014;41(2):43–47; Podsakoff PM, Bommer WH, Podsakoff NP, Mackenzie SB. Relationships between leader reward and punishment behavior and subordinate attitudes, perceptions, and behaviors: a meta-analytic review of existing and new research. Organizational Behavior and Human Decision Processes. 2006; 99(2):113–42; Podsakoff NP, Podsakoff PM, Kuskova VV. Dispelling misconceptions and providing guidelines for leader reward and punishment behavior. Business Horizons. 2010;53(3):291–303.

15. Wang DV, Tsien JZ. Convergent processing of both positive and negative motivational signals by the VTA dopamine neuronal populations. (Research article). PLoS One. 2011;6(2):e17047; Kubanek J, Snyder LH, Abrams RA. Reward and punishment act as distinct factors in guiding behavior. Cognition. 2015;139:154–67.

16. Sidarta A, Vahdat S, Bernardi NF, Ostry DJ. Somatic and reinforcement-based plasticity in the initial stages of human motor learning. Journal of Neuroscience. 2016;36(46):11682; Smith, Graybiel. Habit formation coincides with shifts in reinforcement representations.

17. Fuhrmann A, Kuhl J. Maintaining a healthy diet: effects of personality and self-reward versus self-punishment on commitment to and enactment of self-chosen and assigned goals. Psychology & Health. 1998;13(4):651–86.

18. Greaves CJ, Sheppard KE, Abraham C, et al. Systematic review of reviews of intervention components associated with increased effectiveness in dietary and physical activity interventions. BMC Public Health. 2011;11:119.

19. Specter A, Gerberding JL, Ornish D, Perelson G, et al. Improving nutrition and health through lifestyle modifications: hearing before a subcommittee of the Committee on Appropriations, United States Senate, One Hundred Eighth Congress, first session, special hearing, February 17, 2003, San Francisco, CA: Hearing before the Subcommittee on Departments of Labor, Health, Human Services, Education and Related Agencies (February 17, 2003, 2004); West DS, DiLillo V, Bursac Z, et al. Motivational interviewing improves weight loss in women with type 2 diabetes. (Clinical Care / Education / Nutrition). Diabetes Care. 2007;30(5):1081; Annesi JJ, Vaughn LL. Directionality in the relationship of self-regulation, self-efficacy, and mood changes in facilitating improved physical activity and nutrition behaviors: extending behavioral theory to improve weight-loss treatment effects. Journal of Nutrition Education and Behavior. 2017;49(6):505–12; Greaves, Sheppard, Abraham, et al. Systematic review of reviews of intervention components.

20. Fladung A-K, Gron G, Grammer K, et al. A neural signature of anorexia nervosa in the ventral striatal reward system. American Journal of Psychiatry. 2010;167(2):206.

21. Shea A. How writing a lullaby helps struggling mothers-to-be bond with their babies: WBUR Radio, Boston; 2018. [Available from: http://www.wbur.org/artery/2018/01/25/lullaby-project.]; Wolf DP. Lullaby—Being Together—Being Well. New York: Carnegie Hall's Weill Music Institute; 2017; Hinesley J, Cunningham S, Charles R, et al. The Lullaby Project: a musical intervention for pregnant women. Women's Health Reports. 2020;1(1):543–49.

22. Shea. How writing a lullaby helps struggling mothers-to-be.

23. Hinesley, Cunningham, Charles, et al. The Lullaby Project.

24. Ersche K, Gillan C, Jones P, et al. Carrots and sticks fail to change behavior in cocaine addiction. Science. 2016;352(6292):1468–71.

25. Fairhead J. The significance of death, funerals and the after-life in Ebola-hit Sierra Leone, Guinea and Liberia: anthropological insights into infection and social resistance. Health & Education Advice and Resource Team (HEART); 2014. [Available from: http://www.heart-resources.org/doc_lib/significance-death-funerals-life-ebola-hit-sierra-leone-guinea-liberia-anthropological-insights-infection-social-resistance/.]

26. Abramowitz SA, McLean KE, McKune SL, et al. Community-centered responses to Ebola in urban Liberia: the view from below. PLoS Neglected Tropical Diseases. 2015;9(4):e0003706.

27. Fairhead. The significance of death, funerals and the after-life.

28. Chan M. Ebola virus disease in West Africa—no early end to the outbreak. New England Journal of Medicine. 2014;371(13):1183–85; Yamanis T, Nolan E, Shepler S. Fears and misperceptions of the Ebola response system during the 2014–2015 outbreak in Sierra Leone. (Report). PLoS Neglected Tropical Diseases. 2016;10(10): e0005077.

29. Abramowitz S, McKune SL, Fallah M, et al. The opposite of denial: social learning at the onset of the Ebola emergency in Liberia. Journal of Health Communication.

2017;22:59–65; Abramowitz, McLean, McKune, et al. Community-centered responses to Ebola in urban Liberia; Gamma AE, Slekiene J, Von Medeazza G, et al. Contextual and psychosocial factors predicting Ebola prevention behaviours using the RANAS approach to behaviour change in Guinea-Bissau. BMC Public Health. 2017;17(1):446.

30. Reaves EJ, Mabande LG, Thoroughman DA, et al. Control of Ebola virus disease—Firestone District, Liberia, 2014. Morbidity and Mortality Weekly Report. 2014;63:959; Jalloh MF, Bunnell R, Robinson S, et al. Assessments of Ebola knowledge, attitudes and practices in Forécariah, Guinea and Kambia, Sierra Leone, July–August 2015. Philosophical Transactions of the Royal Society of London Series B, Biological Sciences. 2017;372(1721):20160304.
31. Abramowitz, McLean, McKune, et al. Community-centered responses to Ebola in urban Liberia, 11.
32. Nelson JK, Zeckhauser R. The patron's payoff: conspicuous commissions in Italian Renaissance art. Princeton, NJ: Princeton University Press; 2008.
33. Thaler RH, Sunstein CR. Nudge: improving decisions about health, wealth, and happiness. New Haven, CT: Yale University Press; 2008.
34. Marchiori DR, Adriaanse MA, De Ridder DTD. Unresolved questions in nudging research: putting the psychology back in nudging. Social and Personality Psychology Compass. 2017;11(1):e12297.
35. Loewenstein G, Brennan T, Volpp KG. Asymmetric paternalism to improve health behaviors. JAMA. 2007;298(20):2415–17.
36. Marchiori, Adriaanse, De Ridder. Unresolved questions in nudging research.
37. Dreibelbis R, Kroeger A, Hossain K, et al. Behavior change without behavior change communication: nudging handwashing among primary school students in Bangladesh. International Journal of Environmental Research and Public Health. 2016; 13(1):129; Thedell T. Nudging toward safety progress. Professional Safety. 2016; 61(9):27.
38. Bucher T, Collins C, Rollo ME, et al. Nudging consumers towards healthier choices: a systematic review of positional influences on food choice. British Journal of Nutrition. 2016;115(12):2252–63; Olstad DL, Goonewardene LA, McCargar LJ, Raine KD. Choosing healthier foods in recreational sports settings: a mixed methods investigation of the impact of nudging and an economic incentive. International Journal of Behavioral Nutrition and Physical Activity. 2014;11(1):835; Friis R, Skov LR, Olsen A, et al. Comparison of three nudge interventions (priming, default option, and perceived variety) to promote vegetable consumption in a self-service buffet setting. PLoS One. 12(5):e0176028.
39. Baldwin R. From regulation to behaviour change: giving nudge the third degree. Modern Law Review. 2014;77(6):831–57; Goodwin T. Why we should reject 'nudge.' Politics. 2012;32(2):85–92; Evans N. A 'nudge' in the wrong direction. (Review). Institute of Public Affairs. 2012;64(4):16–9.

40. Hansen P, Jespersen A. Nudge and the manipulation of choice. European Journal of Risk Regulation. 2013;4(1):3–28; Goodwin. Why we should reject 'nudge.'; Evans. A 'nudge' in the wrong direction; Mols F, Haslam SA, Jetten J, Steffens NK. Why a nudge is not enough: a social identity critique of governance by stealth. European Journal of Political Research. 2015;54(1):81–98; Moseley A, Stoker G. Nudging citizens? Prospects and pitfalls confronting a new heuristic. Resources, Conservation & Recycling. 2013;79:4–10.

41. Hansen PG, Skov LR, Skov KL. Making healthy choices easier: regulation versus nudging. Annual Review of Public Health. 2016;37:237–51.

42. Nørnberg TR, Houlby L, Skov LR, Peréz-Cueto FJA. Choice architecture interventions for increased vegetable intake and behaviour change in a school setting: a systematic review. Perspectives in Public Health. 2016;136(3):132–42.

43. Borovoy A, Roberto C. Japanese and American public health approaches to preventing population weight gain: a role for paternalism? Social Science & Medicine. 2015; 143:62.

44. Nørnberg, Houlby, Skov, Peréz-Cueto. Choice architecture interventions for increased vegetable intake and behaviour change; Guthrie J, Mancino L, Lin CTJ. Nudging consumers toward better food choices: policy approaches to changing food consumption behaviors. Psychology & Marketing. 2015;32(5):501–11; Marchiori, Adriaanse, De Ridder. Unresolved questions in nudging research.

8. STRATEGIES FOR PRO-ENVIRONMENTAL SHIFTS

1. Carson R. Silent spring. First Mariner Books edition. Lear LJ, Wilson EO, editors. Boston: Houghton Mifflin Harcourt; 1962 (reissued 2002).

2. Ehrlich PR. The population bomb. New York: Ballantine Books; 1968.

3. Gifford R. Environmental psychology matters. Annual Review of Psychology. 2014;65:541–79; Lehman PK, Gellman ES. Behavior analysis and environmental protection: accomplishments and potential for more. Behavior and Social Issues. 2004;13:13–32.

4. Sundstrom E, Bell P, Busby P, Asmus C. Environmental psychology 1989–1994. Annual Review of Psychology. 1996;47:485; Dwyer WO, Leeming FC, Cobern MK, et al. Critical review of behavioral interventions to preserve the environment. Research since 1980. Environment and Behavior. 1992;25(3):275–321.

5. Kazdin AE. Psychological science's contributions to a sustainable environment. American Psychologist. 2009;64(5):339–56; Raskin PD. World lines: a framework for exploring global pathways. Ecological Economics. 2008;65(3):461–70.

6. Capstick S. Public understanding of climate change as a social dilemma. Sustainability. 2013;5(8):3484–501.

7. Dwyer, Leeming, Cobern, et al. Critical review of behavioral interventions; Sundstrom, Bell, Busby, Asmus. Environmental psychology 1989–1994; Lehman, Gellman.

Behavior analysis and environmental protection; Berthoû S. The everyday challenges of pro-environmental practices. Journal of Transdisciplinary Environmental Studies. 2013;12(1):53–68; Gifford R, Nilsson A. Personal and social factors that influence pro-environmental concern and behaviour: a review. International Journal of Psychology. 2014;49(3):141–57; Raskin. World lines; Stephenson J, Crane SF, Levy C, Maslin M. Population, development, and climate change: links and effects on human health. The Lancet. 2013;382(9905):1665–73.

8. Dietz T, Gardner GT, Gilligan J, et al. Household actions can provide a behavioral wedge to rapidly reduce US carbon emissions. Proceedings of the National Academy of Sciences of the United States of America. 2009;106(44):18452; Pacala S, Socolow R. Stabilization wedges: solving the climate problem for the next 50 years with current technologies. Science. 2004;305(5686):968–72.
9. Kazdin. Psychological science's contributions to a sustainable environment.
10. Berthoû. Everyday challenges of pro-environmental practices.
11. Tybur JM, Griskevicius V. Evolutionary psychology: a fresh perspective for understanding and changing problematic behavior. Public Administration Review. 2013;73(1):12–22.
12. Pfautsch S, Gray T. Low factual understanding and high anxiety about climate warming impedes university students to become sustainability stewards. International Journal of Sustainability in Higher Education. 2017;18(7):1157–75.
13. Griskevicius V, Kenrick DT. Fundamental motives: how evolutionary needs influence consumer behavior. Journal of Consumer Psychology. 2013;23(3):372–86.
14. Tukker A, Cohen MJ, Hubacek K, Mont O. The impacts of household consumption and options for change. Journal of Industrial Ecology. 2010;14(1):13–30.
15. Gifford, Nilsson. Personal and social factors that influence pro-environmental concern and behaviour; Pongiglione F. Motivation for adopting pro-environmental behaviors: the role of social context. Ethics, Policy & Environment. 2014;17(3): 308–23.
16. Gifford, Nilsson. Personal and social factors that influence pro-environmental concern and behaviour; Rezvani Z, Jansson J, Bengtsson M. Consumer motivations for sustainable consumption: the interaction of gain, normative and hedonic motivations on electric vehicle adoption. Business strategy and the environment. 2018; 27(8): 1272–83; Pongiglione. Motivation for adopting pro-environmental behaviors.
17. Akil H, Bouillé J, Robert-Demontrond P. Visual representations of climate change and individual decarbonisation project: an exploratory study. Revue de l'Organisation Responsable. 2017;12(1):66–80.
18. Ryghaug M. Obstacles to sustainable development: the destabilisation of climate change knowledge. Sustainable Development. 2011;19:157–66; Pongiglione F, Cherlet J. The social and behavioral dimensions of climate change: fundamental but disregarded? Journal for General Philosophy of Science. 2015;46(2):383–91.

19. Dietz, Gardner, Gilligan, et al. Household actions can provide a behavioral wedge to rapidly reduce US carbon emissions.

20. Koletsou A, Mancy R. Which efficacy constructs for large-scale social dilemma problems? Individual and collective forms of efficacy and outcome expectancies in the context of climate change mitigation. Risk Management. 2011;13(4):184–208.

21. Pacala, Socolow. Stabilization wedges. Science. 2004;305(5686):968–72.

22. Schwerhoff G, Nguyen T, Edenhofer O, et al. Policy options for a socially balanced climate policy. Economics. 2017;11(20):1–12; Watts N, Adger WN, Agnolucci P, et al. Health and climate change: policy responses to protect public health. The Lancet. 2015;386(10006):1861–914; Fudge S, Peters M. Behaviour change in the UK climate debate: an assessment of responsibility, agency and political dimensions. Sustainability. 2011;3(6):789–808.

23. Shakhashiri B, Bell J. Climate change conversations. Science. 2013;340(6128):9; Rees WE. Human nature, eco-footprints and environmental injustice. Local Environment. 2008;13(8):685–701; Reese G. Common human identity and the path to global climate justice. Climatic Change. 2016;134(4):521–31; Clayton S, Devine-Wright P, Stern P, et al. Psychological research and global climate change. Nature Climate Change. 2015;5(7): 640–46.

24. Decanio S, Fremstad A. Game theory and climate diplomacy. Ecological Economics. 2013;85:177–87; Madani K. Modeling international climate change negotiations more responsibly: can highly simplified game theory models provide reliable policy insights? Ecological Economics. 2013;90:68–76; Soroos M. Global change, environmental security, and the Prisoner's Dilemma. Journal of Peace Research. 1994;31:317; Safarzyńska K, van Den Bergh JCJM. Evolving power and environmental policy: explaining institutional change with group selection. Ecological Economics. 2010;69(4):743–52.

25. Pazzanese C. How Earth Day gave birth to the environmental movement. Harvard Gazette. April 17, 2020.

26. Kirton J. Consequences of the 2008 US elections for America's climate change policy, Canada, and the world. International Journal. 2009;64(1):153–62; Stevens B. Politics, elections and climate change. Social Alternatives. 2007;26(4):10–5.

27. McCrea R, Leviston Z, Walker IA. Climate change skepticism and voting behavior: what causes what? Environment and Behavior. 2016;48(10):1309–34; Tobler C, Visschers VHM, Siegrist M. Addressing climate change: determinants of consumers' willingness to act and to support policy measures. Journal of Environmental Psychology. 2012;32(3):197–207; Clayton, Devine-Wright, Stern, et al. Psychological research and global climate change; Urban J. Are we measuring concern about global climate change correctly? Testing a novel measurement approach with the data from 28 countries. Climatic Change. 2016;139(3–4):397–411.

28. Hornsey M, Harris E, Bain P, Fielding K. Meta-analyses of the determinants and outcomes of belief in climate change. Nature Climate Change. 2016;6(6):622–26.

29. McCrea, Leviston, Walker. Climate change skepticism and voting behavior.
30. Obradovich N. Climate change may speed democratic turnover. Climatic Change. 2017;140(2):135–47.
31. Arroyo V. Are there winning strategies for enacting climate policy? Climate clever: how governments can tackle climate change (and still win elections). Climate Policy. 2013;13(1):142–44; Brody S, Grover H, Vedlitz A. Examining the willingness of Americans to alter behaviour to mitigate climate change. Climate Policy. 2012;12(1): 1–22; Schnoor JL. Coalitions of the willing. Environmental Science & Technology. 2012;46(17):9201.
32. Milfont T, Evans L, Sibley C, et al. Proximity to coast is linked to climate change belief. PLoS One. 2014;9(7):e103180; Clayton, Devine-Wright, Stern, et al. Psychological research and global climate change; Koerth J, Vafeidis A, Hinkel J, Sterr H. What motivates coastal households to adapt pro-actively to sea-level rise and increasing flood risk? Regional Environmental Change. 2013;13(4):897–909.
33. Rosentrater L, Saelensminde I, Ekstrom F, et al. Efficacy trade-offs in individuals' support for climate change policies. Environment and Behavior. 2013;45(8):935–70; Bertolotti M, Catellani P. Effects of message framing in policy communication on climate change. European Journal of Social Psychology. 2014;44(5):474–86.
34. Levin IP, Gaeth GJ. How consumers are affected by the framing of attribute information before and after consuming the product. Journal of Consumer Research. 1988; 15(3):374–78.
35. Hardisty D, Johnson E, Weber E. A dirty word or a dirty world?: Attribute framing, political affiliation, and query theory. Psychological Science. 2010;21(1):86.
36. Gillingham K, Sweeney J. Barriers to implementing low-carbon technologies. Climate Change Economics. 2012;3(04):1250019.
37. Gillingham, Sweeney. Barriers to implementing low-carbon technologies.
38. Gillingham, Sweeney. Barriers to implementing low-carbon technologies.
39. Gillingham, Sweeney. Barriers to implementing low-carbon technologies.
40. Milne MJ, Gray R. W(h)ither ecology? The triple bottom line, the global reporting initiative, and corporate sustainability reporting. Journal of Business Ethics. 2013;118(1):13–29.
41. González-Rodríguez MR, Díaz-Fernández MC, Simonetti B. The social, economic and environmental dimensions of corporate social responsibility: The role played by consumers and potential entrepreneurs. International Business Review. 2015;24(5):836–48; Monast JJ, Adair SK. A triple bottom line for electric utility regulation: aligning state-level energy, environmental, and consumer protection goals. Columbia Journal of Environmental Law. 2013;38(1):1–65; Cervellon M-C. Victoria's dirty secrets: effectiveness of green not-for-profit messages targeting brands. Journal of Advertising. 2012;41(4):133–45; Frederking L. Getting to green: niche-driven or government-led entrepreneurship and sustainability in the wine industry. New England Journal of

Entrepreneurship. 2011;14(1):47–60; Longoni A, Cagliano R. Environmental and social sustainability priorities. International Journal of Operations & Production Management. 2015;35(2):216–45; Brannan DB, Heeter J, Bird L. Made with renewable energy: how and why companies are labeling consumer products. Technical Report NREL / TP-6A20-53764.

Golden, CO: United States Department of Energy National Renewable Energy Laboratory; 2012; Kastner I, Matthies E. Motivation and impact: implications of a twofold perspective on sustainable consumption for intervention programs and evaluation sesigns. Gaia. 2014;23(S1):175–83.

42. Davies ZG, Armsworth PR. Making an impact: the influence of policies to reduce emissions from aviation on the business travel patterns of individual corporations. Energy Policy. 2010;38(12):7634–38; Girod B, van Vuuren DP, Deetman S. Global travel within the 2°C climate target. Energy Policy. 2012;45:152–66; Howitt OJA, Revol VGN, Smith IJ, Rodger CJ. Carbon emissions from international cruise ship passengers' travel to and from New Zealand. Energy Policy. 2010;38(5):2552–60; Williams V, Noland RB, Toumi R. Reducing the climate change impacts of aviation by restricting cruise altitudes. Transportation Research D. 2002;7(6):451–64; Edwards HA, Dixon-Hardy D, Wadud Z. Aircraft cost index and the future of carbon emissions from air travel. Applied Energy. 2016;164:553–62.

43. Dragomir VD. Environmental performance and responsible corporate governance: an empirical note. E & M Ekonomie A Management. 2013;16(1):33–51.

44. Eden S. Environmental issues: nature versus the environment? Progress in Human Geography. 2001;25(1):79–85.

45. Correia D. Degrowth, American style: no impact man and bourgeois primitivism. Capitalism Nature Socialism. 2012;23(1):105–18; Jackson T. Prosperity without growth: economics for a finite planet. London: Earthscan; 2009; Schwerhoff, Nguyen, Edenhofer, et al. Policy options for a socially balanced climate policy.

46. Correia. Degrowth, American style.

47. Jackson. Prosperity without growth; Daly H, editor. A steady state economy. Sustainable Development Commission. London: Earthscan; 2008.

48. Kolbert E. Green like me: living without a fridge, and other experiments in environmentalism. The New Yorker. 2009 August 31; Stern PC. New environmental theories: toward a coherent theory of environmentally significant behavior. Journal of Social Issues. 2000;56(3):407–24.

49. Oreskes N, Conway EM. The collapse of Western civilization: a view from the future. New York: Columbia University Press; 2014.

50. Hornsey M, Harris E, Bain P, Fielding K. Meta-analyses of the determinants and outcomes of belief in climate change. Nature Climate Change. 2016; 6(6):622–26; Kazdin. Psychological science's contributions to a sustainable environment.

51. Di Cosmo V, O'Hora D. Nudging electricity consumption using TOU pricing and feedback: evidence from Irish households. Journal of Economic Psychology. 2017;61:1–14.

52. Staats H, Harland P, Wilke HAM. Effecting durable change: a team approach to improve environmental behavior in the household. (Author abstract). Environment and Behavior. 2004;36(3):341.

53. Kahneman D. Thinking, fast and slow. First ed. New York: Farrar, Straus and Giroux; 2011; Stoknes PE. What we think about when we try not to think about global warming: toward a new psychology of climate action. White River Junction, VT: Chelsea Green Publishing; 2015.

54. Allcott H, Mullainathan S. Energy. Behavior and energy policy. Science. 2010;327(5970):1204; Girod B, van Vuuren DP, Hertwich EG. Climate policy through changing consumption choices: options and obstacles for reducing greenhouse gas emissions. Global Environmental Change. 2014;25(March 2014):5–15.

55. Wilson C, Crane L, Chryssochoidis G. Why do homeowners renovate energy efficiently? Contrasting perspectives and implications for policy. Energy Research & Social Science. 2015;7:12–22.

56. McKibben B. Power to the people: why the rise of green energy makes utility companies nervous. The New Yorker. 2015;91(18):30; Staats, Harland, Wilke. Effecting durable change.

57. Wilson, Crane, Chryssochoidis. Why do homeowners renovate energy efficiently?; Fudge, Peters. Behaviour change in the UK climate debate.

58. Wilson, Crane, Chryssochoidis. Why do homeowners renovate energy efficiently?

59. Devine-Wright P. Think global, act local? The relevance of place attachments and place identities in a climate changed world. Global Environmental Change. 2013;23(1):61–69.

60. Wilson, Crane, Chryssochoidis. Why do homeowners renovate energy efficiently?

61. Stern PC. Psychology and the science of human–environment interactions. American Psychologist. 2000;55(5):523–30; Clark CF, Kotchen MJ, Moore MR. Internal and external influences on pro-environmental behavior: participation in a green electricity program. Journal of Environmental Psychology. 2003;23(3):237–46; Gifford, Nilsson. Personal and social factors that influence pro-environmental concern and behaviour; Gifford. Environmental psychology matters; Whitmarsh L. Behavioural responses to climate change: asymmetry of intentions and impacts. Journal of Environmental Psychology. 2009;29(1):13–23.

62. Moser S, Kleinhückelkotten S. Good intents, but low impacts: diverging importance of motivational and socioeconomic determinants explaining pro-environmental behavior, energy use, and carbon footprint. Environment and Behavior. 2018;50(6):626–56.

63. Ortega-Egea J, García-de-Frutos N, Antolín-López R. Why do some people do "more" to mitigate climate change than others? Exploring heterogeneity in psycho-social associations. PLoS One. 2014;9(9):e106645.

64. Milfont, Evans, Sibley, et al. Proximity to coast is linked to climate change belief; Koerth, Vafeidis, Hinkel, Sterr. What motivates coastal households to adapt pro-actively to sea-level rise?

65. Adams M. Ecological crisis, sustainability and the psychosocial subject: beyond behaviour change. London: Macmillan; 2016, pp. 1–11; Hornsey, Harris, Bain, Fielding. Meta-analyses of the determinants and outcomes of belief in climate change; Chen A, Gifford R. "I wanted to cooperate, but . . .": justifying suboptimal cooperation in a commons dilemma. Canadian Journal of Behavioural Science. 2015;47(4): 282–91.
66. Gifford, Nilsson. Personal and social factors that influence pro-environmental concern and behaviour; Berthoû. Everyday challenges of pro-environmental practices; Biggar M, Ardoin NM. More than good intentions: the role of conditions in personal transportation behaviour. Local Environment. 2017;22(2):141–55; Lin S-P. The gap between global issues and personal behaviors: pro-environmental behaviors of citizens toward climate change in Kaohsiung, Taiwan. Mitigation and Adaptation Strategies for Global Change. 2013;18(6):773–83.
67. Lehman, Gellman. Behavior analysis and environmental protection; Stern. Psychology and the science of human–environment interactions; Whitmarsh. Behavioural responses to climate change.
68. Klöckner CA. The psychology of pro-environmental communication: beyond standard information strategies. London: Macmillan; 2015; Dicaglio J, Barlow KM, Johnson JS. Rhetorical recommendations built on ecological experience: a reassessment of the challenge of environmental communication. Environmental Communication. 2017:1–13; Delmas MA, Fischlein M, Asensio OI. Information strategies and energy conservation behavior: a meta-analysis of experimental studies from 1975 to 2012. Energy Policy. 2013;61:729–39.
69. McKibben B. The end of nature. 1st edition. New York: Random House; 1989.
70. McKibben B. The end of nature. Westminster: Random House Publishing Group; 2014.
71. Tarbell IM. The history of the Standard Oil Company. New York: McClure, Phillips & Company; 1904, p. 1103.
72. Sinclair U. The Jungle: with the author's 1946 introduction. Cambridge, MA: Robert Bentley; 1906 / 1970.
73. Oreskes N, Conway EM. Merchants of doubt: how a handful of scientists obscured the truth on issues from tobacco smoke to global warming. 1st US ed. Conway EM, editor. New York: Bloomsbury Press; 2010.
74. Duan R, Zwickle A, Takahashi B. A construal-level perspective of climate change images in US newspapers. Climatic Change. 2017;142(3–4):345–60; Yang ZJ, Seo M, Rickard LN, Harrison TM. Information sufficiency and attribution of responsibility: predicting support for climate change policy and pro-environmental behavior. Journal of Risk Research. 2014; 18(6):727–46; Ramkissoon H, Smith L. The relationship between environmental worldviews, emotions and personal efficacy in climate change. International Journal of Arts & Sciences. 2014;7(1):93–109.
75. O'Neill SJ, Boykoff M, Niemeyer S, Day SA. On the use of imagery for climate change engagement. Global Environmental Change. 2013;23(2):413–21.

76. O'Neill SJ, Hulme M. An iconic approach for representing climate change. Global Environmental Change. 2009;19(4):402–10.
77. Geiger N, Bowman C, Clouthier T, et al. Observing environmental destruction stimulates neural activation in networks associated with empathic responses. Social Justice Research. 2017;30(4):300–22.
78. Meijnders AL, Midden CJH, Wilke HAM. Communications about environmental risks and risk-reducing behavior: the impact of fear on information processing. Journal of Applied Social Psychology. 2001;31(4):754–77.
79. Klöckner. The psychology of pro-environmental communication.
80. Howell RA. Lights, camera . . . action? Altered attitudes and behaviour in response to the climate change film The Age of Stupid. Global Environmental Change. 2011;21(1):177–87; Klöckner. The psychology of pro-environmental communication; Tsitsoni V, Toma L. An econometric analysis of determinants of climate change attitudes and behaviour in Greece and Great Britain. Agricultural Economics Review. 2013;14(1):59–75; Ramkissoon, Smith. The relationship between environmental worldviews; Maibach E. Social marketing for the environment: using information campaigns to promote environmental awareness and behavior change. Health Promotion International. 1993;8(3):209–24.
81. Spartz JT, Su LY-F, Griffin R, et al. YouTube, social norms and perceived salience of climate change in the American mind. Environmental Communication. 2017;11(1): 1–16; van der Linden S, Leiserowitz A, Mailbach E. The gateway belief model: a large-scale replication. Journal of Environmental Psychology. 2019;62(4):49–58.
82. Dunlap RE. Climate change skepticism and denial. American Behavioral Scientist. 2013;57(6):691–98; Oreskes N, Conway EM. Defeating the merchants of doubt. Nature. 2010;465:686–87; Skoglund A, Stripple J. From climate skeptic to climate cynic. Critical Policy Studies. 2018;13(3):345–65; Stern P, Perkins J, Sparks R, Knox R. The challenge of climate-change neoskepticism. Science. 2016;353(6300):653–54.
83. Brüggemann M, Engesser S. Beyond false balance: how interpretive journalism shapes media coverage of climate change. Global Environmental Change. 2017;42:58–67; Ryghaug. Obstacles to sustainable development.
84. Brüggemann, Engesser. Beyond false balance; Bushell S, Buisson GS, Workman M, Colley T. Strategic narratives in climate change: towards a unifying narrative to address the action gap on climate change. Energy Research & Social Science. 2017;28:39–49; Hall C. Framing behavioural approaches to understanding and governing sustainable tourism consumption: beyond neoliberalism, "nudging" and "green growth"? Journal of Sustainable Tourism. 2013;21(7):1091–1109.
85. Bertolotti, Catellani. Effects of message framing in policy communication on climate change; Mailbach EW, Nisbet M, Baldwin P, et al. Reframing climate change as a public health issue: an exploratory study of public reactions. BMC Public Health. 2010;10(1):299.
86. Stoknes. What we think about when we try not to think about global warming.

87. Schubert C. Green nudges: do they work? Are they ethical? Ecological Economics. 2017;132:329–42.

88. Adams. Ecological crisis, sustainability and the psychosocial subject; Ryghaug. Obstacles to sustainable development.

89. Dicaglio, Barlow, Johnson. Rhetorical recommendations built on ecological experience.

90. Huang H. Media use, environmental beliefs, self-efficacy, and pro-environmental behavior. Journal of Business Research. 2016;69(6):2206–12; O'Neill, Boykoff, Niemeyer, Day. On the use of imagery for climate change engagement; Hart PS, Feldman L. The influence of climate change efficacy messages and efficacy beliefs on intended political participation. (Research article). PLoS One. 2016;11(8):e0157658; Tsitsoni, Toma. An econometric analysis of determinants of climate change attitudes.

91. Markowitz E, Shariff A. Climate change and moral judgement. Nature Climate Change. 2012;2(4):243–47; Markowitz E, Slovic P, Västfjäll D, Hodges S. Compassion fade and the challenge of environmental conservation. Judgment and Decision Making. 2013;8(4):397.

92. Gifford R. The dragons of inaction: psychological barriers that limit climate change mitigation and adaptation. American Psychologist. 2011;66(4):290–302; Tobler, Visschers, Siegrist. Addressing climate change; Swim J, Clayton S, Doherty T, et al. Psychology and global climate change: addressing a multi-faceted phenomenon and set of challenges. American Psychological Association; 2011. [Available at: http://www.apa.org/science/about/publications/climate-change.aspx]; Schutte NS, Bhullar N. Approaching environmental sustainability: perceptions of self-efficacy and changeability. Journal of Psychology. 2017;151(3):321–33; Huang. Media use, environmental beliefs, self-efficacy, and pro-environmental behavior; Hart, Feldman. The influence of climate change efficacy messages; Koletsou, Mancy. Which efficacy constructs for large-scale social dilemma problems?; Roser-Renouf C, Mailbach EW, Leiserowitz A, Zhao X. The genesis of climate change activism: from key beliefs to political action. Climatic Change 2014;125(2):163–78.

93. Bissing-Olson MJ, Fielding KS, Iyer A. Experiences of pride, not guilt, predict pro-environmental behavior when pro-environmental descriptive norms are more positive. Journal of Environmental Psychology. 2016;45:145–53; Chamila Roshani Perera L, Rathnasiri Hewege C. Climate change risk perceptions and environmentally conscious behaviour among young environmentalists in Australia. Young Consumers. 2013;14(2):139–54.

94. Staats, Harland, Wilke. Effecting durable change.

95. Fisher J, Irvine K. Reducing energy use and carbon emissions: a critical assessment of small-group interventions. Energies. 2016;9:172.

96. Staats, Harland, Wilke. Effecting durable change; Fisher, Irvine. Reducing energy use and carbon emissions.

97. Klöckner. The psychology of pro-environmental communication.

98. Lokhorst AM, Werner C, Staats H, et al. Commitment and behavior change. Environment and Behavior. 2013;45(1):3–34.

99. Kruijsen JHJ, Owen A, Boyd DMG. Community sustainability plans to enable change towards sustainable practice—a Scottish case study. Local Environment. 2014;19(7):748–66.

100. Mols F, Haslam SA, Jetten J, Steffens NK. Why a nudge is not enough: a social identity critique of governance by stealth. European Journal of Political Research. 2015;54(1): 81–98.

101. Schwerhoff, Nguyen, Edenhofer, et al. Policy options for a socially balanced climate policy; Bristow AL, Wardman M, Zanni AM, Chintakayala PK. Public acceptability of personal carbon trading and carbon tax. Ecological Economics. 2010;69(9):1824–37; Murray B, Rivers N. British Columbia's revenue-neutral carbon tax: a review of the latest "grand experiment" in environmental policy. Energy Policy. 2015;86(C):674–83; Schiermeier Q. Anger as Australia dumps carbon tax. Nature. 2014;511(7510):392; Parag Y, Capstick S, Poortinga W. Policy attribute framing: a comparison between three policy instruments for personal emissions reduction. Journal of Policy Analysis and Management. 2011;30(4):889–905.

102. Guzman LI, Clapp A. Applying personal carbon trading: a proposed 'Carbon, Health and Savings System' for British Columbia, Canada. Climate Policy. 2017; 17(5):616–33.

103. Hendry A, Webb G, Wilson A, et al. Influences on intentions to use a personal carbon trading system (NICHE—The Norfolk Island Carbon Health Evaluation Project). International Technology Management Review. 2015;5(2):105–16; Guzman, Clapp. Applying personal carbon trading.

104. Stoknes. What we think about when we try not to think about global warming.

105. Gifford. The dragons of inaction; de Nazelle A, Fruin S, Westerdahl D, et al. A travel mode comparison of commuters' exposures to air pollutants in Barcelona. Atmospheric Environment. 2012;59:151–59.

106. Tybur, Griskevicius. Evolutionary psychology.

107. Koletsou, Mancy. Which efficacy constructs for large-scale social dilemma problems?

108. Di Cosmo, O'Hora. Nudging electricity consumption using TOU pricing and feedback.

109. Pichert D, Katsikopoulos KV. Green defaults: information presentation and pro-environmental behaviour. Journal of Environmental Psychology. 2008;28(1):63–73; Ölander F, Thøgersen J. Informing versus nudging in environmental policy. Journal of Consumer Policy. 2014;37(3):341–56; Ebeling F, Lotz S. Domestic uptake of green energy promoted by opt-out tariffs. Nature Climate Change. 2015;5(9):868–71.

110. Graffeo M, Ritov I, Bonini N, Hadjichristidis C. To make people save energy tell them what others do but also who they are: a preliminary study. Frontiers of Psychology. 2015;6:1287; Ölander, Thøgersen. Informing versus nudging in environmental policy.

111. Stea S, Pickering GJ. Optimizing messaging to reduce red meat consumption. Environmental Communication. 2017;13(5):633–48.

112. Barnes AP, Toma L, Willock J, Hall C. Comparing a 'budge' to a 'nudge': farmer responses to voluntary and compulsory compliance in a water quality management regime. Journal of Rural Studies. 2013;32:448–59.
113. Mills J, Gaskell P, Ingram J, et al. Engaging farmers in environmental management through a better understanding of behaviour. Agriculture and Human Values. 2017; 34(2):283–99.
114. Oliver A. From nudging to budging: using behavioural economics to inform public sector policy. Journal of Social Policy. 2013;42(4):685–700; Mols, Haslam, Jetten, Steffens. Why a nudge is not enough; Moseley A, Stoker G. Nudging citizens? Prospects and pitfalls confronting a new heuristic. Resources, Conservation & Recycling. 2013;79:4–10.
115. MacKay DJC. Sustainable energy without the hot air. Cambridge: UIT Cambridge; 2009.
116. Goodwin T. Why we should reject 'nudge.' Politics. 2012;32(2):85–92; Lehner M, Mont O, Heiskanen E. Nudging—a promising tool for sustainable consumption behaviour? Journal of Cleaner Production. 2016;134:166–77.
117. Moseley, Stoker. Nudging citizens? Prospects and pitfalls.
118. Newell PB. A cross-cultural examination of favorite places. Environment and Behavior. 1997;29(4):495.
119. Annerstedt van den Bosch M, Depledge MH. Healthy people with nature in mind. BMC Public Health. 2015;15(1):1232–32; Frumkin H. The evidence of nature and the nature of evidence. American Journal of Preventive Medicine. 2012;44(2):196–97; Shanahan DF, Fuller RA, Bush R, et al. The health benefits of urban nature: how much do we need? BioScience. 2015;65(5):476–85.
120. Uren HV, Dzidic PL, Roberts LD, et al. Green-tinted glasses: how do pro-environmental citizens conceptualize environmental sustainability? Environmental Communication. 2019;13(3):395–411; Eden. Environmental issues.
121. Beery TH, Wolf-Watz D. Nature to place: rethinking the environmental connectedness perspective. Journal of Environmental Psychology. 2014;40:198–205.
122. Eden. Environmental issues.
123. Chawla L, Derr V. The development of conservation behaviors in childhood and youth. In: Clayton SD, editor. The Oxford handbook of environmental and conservation psychology. New York: Oxford University Press; 2012, pp. 1–48.
124. Wells NM, Lekies KS. Nature and the life course: pathways from childhood nature experiences to adult environmentalism. Children Youth and Environments. 2006;16(1): 1–24; Chawla, Derr. The development of conservation behaviors; Broom C. Exploring the relations between childhood experiences in nature and young adults' environmental attitudes and behaviours. Australian Journal of Environmental Education. 2017;33(1):34–47; Klaniecki K, Leventon J, Abson D. Human–nature connectedness as a 'treatment' for pro-environmental behavior: making the case for spatial considerations. Sustainability Science. 2018:1–14.

125. Ward Thompson C, Aspinall P, Montarzino A. The childhood factor: adult visits to green places and the significance of childhood experience. (Report). Environment and Behavior. 2008;40(1):111.
126. Hsu S-J. Significant life experiences affect environmental action: a confirmation study in eastern Taiwan. Environmental Education Research. 2009;15(4):497–517.
127. Ward Thompson, Aspinall, Montarzino. The childhood factor; Chawla, Derr. The development of conservation behavior; Calogiuri G. Natural environments and childhood experiences promoting physical activity, examining the mediational effects of feelings about nature and social networks. International Journal of Environmental Research and Public Health. 2016;13(4); Larson LR, Green GT, Castleberry SB. Construction and validation of an instrument to measure environmental orientations in a diverse group of children. Environment and Behavior. 2011;43(1):72–89.
128. Larson, Green, Castleberry. Construction and validation of an instrument to measure environmental orientations.
129. Collado S, Corraliza JA, Staats H, Ruiz M. Effect of frequency and mode of contact with nature on children's self-reported ecological behaviors. Journal of Environmental Psychology. 2015;41:65–73.
130. Von Lindern E, Bauer N, Frick J, et al. Occupational engagement as a constraint on restoration during leisure time in forest settings. Landscape and Urban Planning. 2013;118(C):90–97.
131. Park JJ, Selman P. Attitudes toward rural landscape change in England. Environment and Behavior. 2011;43(2):182–206.
132. Coldwell D, Evans K. Contrasting effects of visiting urban green-space and the countryside on biodiversity knowledge and conservation support. PLoS One. 2017;12(3):e0174376.
133. Ohly H, Gentry S, Wigglesworth R, et al. A systematic review of the health and well-being impacts of school gardening: synthesis of quantitative and qualitative evidence. (Report). BMC Public Health. 2016;16(1):286.
134. White R, Eberstein K, Scott D. Birds in the playground: evaluating the effectiveness of an urban environmental education project in enhancing school children's awareness, knowledge and attitudes towards local wildlife. PLoS One. 2018;13(3):e0193993; Ali SM, Rostam K, Awang AH. School landscape environments in assisting the learning process and in appreciating the natural environment. Procedia—Social and Behavioral Sciences. 2015;202:189–98; Bell SL, Westley M, Lovell R, Wheeler BW. Everyday green space and experienced well-being: the significance of wildlife encounters. Landscape Research. 2018;43(1):8–19; Otto S, Pensini P. Nature-based environmental education of children: environmental knowledge and connectedness to nature, together, are related to ecological behaviour. Global Environmental Change. 2017;47:88–94.
135. Chawla, Derr. The development of conservation behaviors.

136. Bixler RD, Floyd MF, Hammitt WE. Environmental socialization: quantitative tests of the childhood play hypothesis. Environment and Behavior. 2002;34(6): 795–818.

137. Hsu. Significant life experiences affect environmental action.

138. Huber R, Panksepp JB, Nathaniel T, et al. Drug-sensitive reward in crayfish: an invertebrate model system for the study of SEEKING, reward, addiction, and withdrawal. Neuroscience and Biobehavioral Reviews. 2011;35(9):1847–53.

9. THE GREEN CHILDREN'S HOSPITAL

1. Council on Environmental Health. Global climate change and children's health. Pediatrics. 2015;136(5):992.

2. Dienes C. Actions and intentions to pay for climate change mitigation: environmental concern and the role of economic factors. Ecological Economics. 2015;109:122–29.

3. McKibben B. A special moment in history. Atlantic Monthly. 1998 May;281(5): 55–78.

4. Ramage N. Sustainable marketing. Marketing. 2005;110(9):6; Oreskes N, Conway EM. Defeating the merchants of doubt. Nature. 2010;465:686–87.

5. Chung JW, Meltzer DO. Estimate of the carbon footprint of the US health care sector. JAMA. 2009;302(18):1970.

6. Eckelman MJ, Sherman J. Environmental impacts of the U.S. health care system and effects on public health. (Report). PLoS One. 2016;11(6):e0157014.

7. Eckelman, Sherman. Environmental impacts of the U.S. health care system.

8. Kaplan S, Sadler B, Little K, et al. Can sustainable hospitals help bend the health care cost curve? The Commonwealth Fund. 2012;29:1–14.

9. Howard J, Hill L, Krause D. Defining waste and material streams. Practice Greenhealth; 2015. [Available at: https://practicegreenhealth.org/sites/default/files/upload-files/defining_waste_and_material_streams.pdf.]

10. Riedel LM. Environmental and financial impact of a hospital recycling program. AANA Journal. 2011;79(4 suppl):S8.

11. Veleva V, Bodkin G. Corporate-entrepreneur collaborations to advance a circular economy. Journal of Cleaner Production. 2018;188:20–37; Ferenc J. Model workers. Trustee. 2016;69(6):3; Howard, Hill, Krause. Defining waste and material streams.

12. Babu MA, Dalenberg AK, Goodsell G, et al. Greening the operating room: results of a scalable initiative to reduce waste and recover supply costs. Neurosurgery. 2019;85(3): 432–37.

13. Allen JG, MacNaughton P, Satish U, et al. Associations of cognitive function scores with carbon dioxide, ventilation, and volatile organic compound exposures in office workers: a controlled exposure study of green and conventional office environments. Environmental Health Perspectives. 2015;124(6):805–12.

14. Riesenberg DE, Arehart-Treichel J. "Sick building" syndrome plagues workers, dwellers. JAMA. 1986;255(22):3063; Redlich CA, Sparer J, Cullen MR. Sick-building syndrome. The Lancet. 1997;349(9057):1013–16.

15. Persily A. Challenges in developing ventilation and indoor air quality standards: the story of ASHRAE Standard 62. Building and Environment. 2015;91(C):61–69.

16. Brandt-Rauf PW, Andrews LR, Schwarz-Miller J. Sick-hospital syndrome. Journal of Occupational Medicine: Official Publication of the Industrial Medical Association. 1991;33(6):737; Leung M, Chan AHS. Control and management of hospital indoor air quality. Medical Science Monitor: International Medical Journal of Experimental and Clinical Research. 2006;12(3):SR17.

17. Persily. Challenges in developing ventilation and indoor air quality standards.

18. Allen, MacNaughton, Satish, et al. Associations of cognitive function scores with carbon dioxide.

19. Thiel CL, Needy KL, Ries R, et al. Building design and performance: a comparative longitudinal assessment of a children's hospital. Building and Environment. 2014;78:130–36.

20. Eshel G, Shepon A, Makov T, Milo R. Land, irrigation water, greenhouse gas, and reactive nitrogen burdens of meat, eggs, and dairy production in the United States. Proceedings of the National Academy of Sciences of the United States of America. 2014;111(33):11996–2001; Springmann M, Mason-D'Croz D, Robinson S, et al. Global and regional health effects of future food production under climate change: a modelling study. The Lancet. 2016;387(10031):1937–46; Girod B, van Vuuren DP, Hertwich EG. Climate policy through changing consumption choices: options and obstacles for reducing greenhouse gas emissions. Global Environmental Change. 2014;25(March):5–15.

21. Pan A, Sun Q, Bernstein AM, et al. Red meat consumption and mortality: results from 2 prospective cohort studies. Archives of Internal Medicine. 2012;172(7):555–63; Soret S, Mejia A, Batech M, et al. Climate change mitigation and health effects of varied dietary patterns in real-life settings throughout North America. American Journal of Clinical Nutrition. 2014;100(suppl 1):490S; Nemecek T, Jungbluth N, Canals L, Schenck R. Environmental impacts of food consumption and nutrition: where are we and what is next? International Journal of Life Cycle Assessment. 2016;21(5):607–20; Eshel G, Shepon A, Noor E, Milo R. Environmentally optimal, nutritionally aware beef replacement plant-based diets. Environmental Science & Technology. 2016;50(15):8164–68; Tilman D, Clark M. Global diets link environmental sustainability and human health. Nature. 2014;515(7528):518–22; Reynolds CJ, Buckley JD, Weinstein P, Boland J. Are the dietary guidelines for meat, fat, fruit and vegetable consumption appropriate for environmental sustainability? A review of the literature. Nutrients. 2014;6(6):2251.

22. Hart J. Practice Greenhealth: leading efforts to ensure a sustainable health care system and improved health for all. Alternative & Complementary Therapies. 2022;27(5):250–52.

23. Klein K. Values-based food procurement in hospitals: the role of health care group purchasing organizations. Agriculture and Human Values. 2015;32(4):635–48.
24. Ulrich RS. View through a window may influence recovery from surgery. Science. 1984; 224(4647):420–21; Berry LL, Parker D, Coile RC, et al. The business case for better buildings. Frontiers of Health Services Management. 2004;21(1):3.
25. Berry, Parker, Coile, et al. The business case for better buildings; Bardwell P. A challenging road toward a rewarding destination. Frontiers of Health Services Management. 2004;21(1):27–34.
26. Ulrich RS. Essay: evidence-based health-care architecture. The Lancet. 2006;368(1): S38–S39.
27. Klein. Values-based food procurement in hospitals.
28. DeLind LB. Are local food and the local food movement taking us where we want to go? Or are we hitching our wagons to the wrong stars? Agriculture and Human Values. 2010;28(2):273–83.
29. Kaplan, Sadler, Little, et al. Can sustainable hospitals help bend the health care cost curve?
30. Unger SR, Campion N, Bilec MM, Landis AE. Evaluating quantifiable metrics for hospital green checklists. Journal of Cleaner Production. 2016;127:134–42.
31. Africa J, Logan A, Mitchell R, et al. The Natural Environments Initiative: illustrative review and workshop statement. The Natural Environments Initiative; The Radcliffe Institute for Advanced Study, Harvard University: Harvard School of Public Health, Center for Health and the Global Environment; 2014.

10. CONCLUSION

1. Intergovernmental Panel on Climate Change. IPCC, 2021: Summary for policymakers. Sixth assessment report (AR6) of the Intergovernmental Panel on Climate Change. In: Masson-Delmotte V, Zhai P, Pirani A, et al., editors. Climate change 2021: the physical science basis contribution of Working Group I to the Sixth Assessment Report of the Intergovernmental Panel on Climate Change. Cambridge: Cambridge University Press; 2021 (in press).
2. Periyakoil VS, Neri E, Fong A, Kraemer H. Do Unto Others: Doctors' Personal End-of-Life Resuscitation Preferences and Their Attitudes toward Advance Directives. (Research Article). PLoS ONE. 2014;9(5).; Zhang B, Wright AA, Huskamp HA, Nilsson ME, Maciejewski ML, Earle CC, Block SD, Maciejewski PK, Prigerson HG. Health Care Costs in the Last Week of Life: Associations With End-of-Life Conversations. Archives of Internal Medicine. 2009;169(5):480–8.
3. Gladwell M. The tipping point: how little things can make a big difference. Boston: Back Bay Books; 2002.
4. Vaillant GE. Triumphs of experience: the men of the Harvard Grant Study. Cambridge, MA: Belknap Press of Harvard University Press; 2012.

Acknowledgments

This project represents the sustained effort and support of many people who contributed in numerous ways to its ultimate content and completion over a several-year period. Neurologists Zelime Elibold and Alice Flaherty helped refine the original idea in a series of informal meetings at the project's inception. Special thanks go to Dan Schrag, director of the Harvard University Center for the Environment and a long-view geochemist, who hosted an exploratory seminar, supported the validity of this exploration, and welcomed me into the big tent of the center, from which I continue to learn immensely from real experts. Scientific, interdisciplinary, and inspirational support came from pediatrician Cindy Christian, neurosurgeons Emad Eskandar and Ziv Williams, economist / psychiatrist Niels Rosenquist, psychiatrist Jim Recht, neuroradiologist Paul Caruso, infectious disease and climate expert Regina LaRocque, and nature appreciator Ava McNichol, among others. My neurosurgery colleagues Robert Martuza and Bill Butler graciously accommodated my sabbatical at the Radcliffe Institute for Advanced Study at Harvard University to delve into the subject in earnest, and I am forever grateful to former Radcliffe dean Lizabeth Cohen and the late fellowship director Judy Vichniac, who took a chance by admitting their first neurosurgeon into that program and by supporting this interdisciplinary work.

While I was at Radcliffe, undergraduate research assistants Deng-Tung Wang, Fatuma Rinderknecht, Natalie Cho, Daniel Letchford, and Vibav Mouli provided invaluable help and perspective, and fellow Fellows Robert Huber, Reiko Yamada, Valerie Massadian, Karole Armitage, Philip Klein, Esther Yeger-Lotem,

Wendy Gan, and Raj Pandit, along with many others, were instrumental in widening my view and keeping me grounded. Neuroscientist Peter Sterling was instrumental in helping refine the ideas and in providing critical manuscript feedback. My editor at Harvard University Press, Janice Audet, has served as a patient guide through the peer review and editing process, and the peer reviewers from different fields helped me hone and focus the book, and to make it more of a story. Figure 1 was created by Angelynn Grant based on an initial design by the Radcliffe research assistants, and Figures 2 and 3 were created by Elaine Kurie (Kurie Biomedical Illustrations).

My husband, Stan Pelli, and children, Jonas and Alida Pelli, as well as extended family and friends, provided space, time, freedom, support, and extraordinary patience to allow me to pursue this exploration, and to them and all those above I recognize a debt of gratitude that can never be repaid.

Index

Note: The letter "f" following a page number denotes a figure.